RENOLD LIMITED

BY THE SAME AUTHOR

Grand Alliance
A Chapter of Industrial History

Renold Chains
A History of the Company and the Rise of the Precision Chain Industry 1879–1955

The Joint Iron Council 1945–1966
The Origins and History of the Council and its Constituents

SIR CHARLES RENOLD

RENOLD LIMITED
1956–1967

BY

BASIL H. TRIPP

WITH A PREFACE BY

LORD BOWDEN OF CHESTERFIELD
MA, PhD, MSc Tech, FIEE
Principal, University of Manchester Institute of Science and Technology

LONDON
GEORGE ALLEN & UNWIN LTD
RUSKIN HOUSE · MUSEUM STREET

FIRST PUBLISHED IN 1969

SBN 04 331041 9

PRINTED IN GREAT BRITAIN
in 11 on 13 point Baskerville type
AT THE SHENVAL PRESS
LONDON, HERTFORD AND HARLOW

PREFACE

Historians and observers of contemporary England will get a very distorted picture of our society if they depend entirely on the newspapers, on the proceedings of Parliament, or on the much publicized speeches of politicians. For good or ill, the lives of most people really depend on the change and development of industry.

Furthermore, the process of industrial development, the way in which firms grow, develop and collapse, contains elements of high drama however obscure the details may be to the man in the street; the success of a firm and the prosperity of thousands of people often depend on technical achievements of the greatest ingenuity, many of which are quite outside the mainstream of engineering science and which conventional historians neglect—because they can neither comprehend them nor even describe them.

This volume describes the growth of the Renold organization during the last decade and shows how it has come about that an apparently simple component, the bicycle chain, should have become the basis of a great industry upon which many thousands of people depend for their livelihood. Half-a-dozen other firms have joined forces with the original Renold company and created an industrial complex which has diversified its interests and which plays a vital role in many parts of modern industry in almost every country in the world; here in this book we can learn how the Renold group is preparing to face the future.

How many people realize that motor-car manufacturers from every country in Europe habitually send prototypes of their new engines to be tested in Stockport, so that the chain drive upon which the engine depends can be properly designed and fitted? I know of no other English firm which has monopolized a particular technique so completely as to dominate the markets of half the world in this way.

The firm was originally based upon the pioneering invention of Hans Renold, who survived until a very great age, but for the last thirty years it was dominated by his son, that very great man, Sir Charles Renold, whose life is described at some length in this book. I am proud to remember that he was a friend of mine and I am happy to have this opportunity of adding my own tribute to his memory. For, at a time when he was intensively preoccupied with most complex technical and industrial problems, and with the reorganization of a great industry, he still found time to play a vitally important part in the creation and development of this University Institute of ours which became a second home to him. A country is indeed fortunate when great enterprises are in such competent hands and when industrialists—men upon whom rests the burden of creating the wealth upon which we all live—can find time and energy to devote to public affairs.

I would like to wish the company well. I hope that it prospers and that those who will control the company in future are as fortunate, as skilful and as successful as the team which was dominated for so long by Sir Charles Renold and his father.

VIVIAN BOWDEN

The University of Manchester
Institute of Science and Technology
Manchester
July 30, 1968

AUTHOR'S NOTE AND ACKNOWLEDGMENTS

The first book about the company formerly named Renold Chains appeared in 1956. It was a history of the enterprise and of the rise of the precision chain industry from 1879 to 1955. In the comparatively short period between that volume and this, the speed and extent of Renold Group growth have matched the scale of development in industry generally. Much more has happened much faster than it once did.

Moreover, the past twelve years have witnessed a significant organic change in the Group. It is no longer dedicated wholly and solely to the production of precision chain and chain accessories, but has elected to move into a wider field of positive power transmission which, although complementary to chain, vastly extends the product range.

For these reasons, the late Sir Charles Renold and the Group Board kindly asked me for a second book, telling the story of what had happened in this new phase.

Many people have given me generous help and encouragement, first among them Sir Charles Renold himself, up to the days of his last illness, and then Mr Leslie J. Tolley, the Group Managing Director, together with other past and present directors, managers and members of the staffs of Renold Ltd., its predecessors and associated companies. I am most grateful to them all, and especially to Mr C. D. Hughes, for his invariable patience and efficiency in producing whatever documents or information I needed.

The book was completed before the acquisition by Renold Ltd., in 1967, of Crofts Engineers (Holdings) Ltd. So far as is essential, the text has been amended to take account of this, but it was, and is, confined mainly to the description of events before then.

Mr D. J. Drake, of Stanton and Staveley Ltd., and Mr K. W. Humphreys, Librarian to the University of Birmingham, helped me with information, and I thank them.

B. H. T.

Sanderstead, Surrey
1968

CONTENTS

ILLUSTRATIONS

DRAWINGS AND DIAGRAMS

CHRONOLOGY

Below are given, year by year, the principal events in Renold history, together with those in the companies subsequently merged in the Renold Group. Many of these occurrences received full treatment in the earlier work, *Renold Chains*, published in 1956.

1789 Birth of Isaac Brampton at Leicester.

1796 Birth of James Perry, founder of the Perry business (died 1843).

1824 James Perry founds Perry and Co. at Manchester, later moving to London.

1828 Josiah (later Sir Josiah) Mason begins to make steel pens.

1852 Birth of Hans Renold at Aarau, Switzerland. Frederick Brampton, son of Isaac, moves to Birmingham.

1857 W. E. Wiley (subsequently associated with Perry) patents a propelling pencil.

1862 John Holroyd in business in Manchester as an 'engineer and fitter'.

1864 First patent for chain driving granted to James Slater. Arthur Brampton born.

1867 Birth of Alick Sargeant Hill, subsequently Chairman of the Coventry Chain Co. Ltd. (died 1921).

1869 John Holroyd moves to Hulme, Manchester, and is described as a sewing-machine maker.

1870 Sir Josiah Mason acquires A. Sommerville and Co., steel pen makers, Lancaster Street Works, Birmingham.

1873 Hans Renold comes to England.

1874 Death of Frederick Brampton. His two elder sons, Charles Henry (aged 18) and Frederick William (aged 17), trade as 'F. Brampton and Co.'

1876 Perry and Co. Ltd. established at Lancaster Street Works, merging with A. Sommerville and Co., and Wiley and Son, pen and pencil-case makers.

1879 Hans Renold buys the Slater textile chain-making business, Salford.
J. K. Starley's first 'safety' bicycle.

1880 Hans Renold invents and patents the bush roller chain.
Hans Renold becomes a British subject.

1881 Death of Sir Josiah Mason.

1883 Charles Garonne Renold, eldest son of Hans Renold, born.

c. 1885 The brothers Brampton enter the cycle industry and experiment with a self-lubricating bicycle chain.

1887 Establishment of John Holroyd and Co. Ltd., under the chairmanship of Emil Liebert, merchant, of Manchester.

1889 Hans Renold builds the first of his own factories, Progress Works, Brook Street, Manchester. He invents and patents the inverted-tooth ('silent') chain.

1890 Perry enter the chain and cycle fittings industry.

1892 Mr C. T. Bayliss, younger son of J. W. Bayliss, born.

1896 Holroyd move to the present site at Milnrow, Rochdale.
A. S. Hill founds the Coventry Chain Co.
Hans Renold introduces the 48-hour week in his business.
C. H. and F. W. Brampton take into partnership their younger brothers, Arthur and Walter, trading as 'Brampton Brothers'.

1897 Formation of the public company of Brampton Brothers Ltd.

1898 J. W. Bayliss (died 1951) takes charge of the Perry chain and cycle accessories.
Brampton Brothers open a works at Calais.

1902 John Holroyd and Co. Ltd., a new company, formed to take over the business established in 1887.
Coventry Chain becomes a private limited company.

1903 Hans Renold Ltd. formed as a private limited company, with Hans Renold as Governing Director and Chairman.

1906 Hans Renold plans and begins construction of Renold Works, Burnage, near Manchester.
C. G. Renold appointed a Director of Hans Renold Ltd.

1907 Coventry Chain becomes a public company, and approves plans for new works.
New tooth form for roller chain wheels developed by Hans Renold.
Holroyd acquire the ironfounding company of John Petrie and Co. Ltd., Rochdale.

1909 R. O. Herford apprenticed to Hans Renold Ltd.
Foundation of the Hans Renold Social Union.
Arnold and Stolzenberg GmbH, precision chain manufacturers, established at Juliusmühle, Einbeck, Lower Saxony.

1910 Hans Renold Ltd. establish a Personnel Department.

1912 Mr C. T. Bayliss (aged 20) designs the Perry car.
The Perry Motor Co. Ltd. formed.
Martyn Smith, grand-nephew of Sir Josiah Mason, becomes Chairman of Perry.

1914 H. G. Jenkins joins Hans Renold Ltd.
Occupation of Renold Works completed.
First World War: Hans Renold Ltd. concentrate on munitions production.

1916 Hans Renold Ltd. make the original experiments on tank track chains.
Coventry Chain manufacture tank track chains.
Hans Renold invents and patents the segmental bush bearing.
C. G. Renold appointed Works Director of Hans Renold Ltd. He begins to study constitutional management.

1917 Shop Stewards Committee formed at Renold Works; beginning of the policy of joint consultation.
R. O. Herford appointed a Director of Hans Renold Ltd.

1919 C. G. Renold appointed Managing Director of Hans Renold Ltd.
Mr C. T. Bayliss and A. E. Wiley (a descendant of the Wileys of Wiley and Co., died 1964) acquire the Perry Motor Co., subsequently selling to A. Harper, Sons and Bean Ltd.
Mr C. T. Bayliss and A. E. Wiley in business as Bayliss, Wiley and Co. Ltd., makers of cycle bearing components.

1920 Holroyd establish a bronze foundry (now the Holfos Works) at Rochdale.

1921 Formation of SA des Etablissements Brampton, Calais.

1922 Adoption of profit-sharing scheme by Hans Renold Ltd.

1923 Mr O. Bertoya joins Hans Renold Ltd.

1925 Merger of Brampton Brothers and Coventry Chain.

1926 Bayliss, Wiley merge with Perry.

1927 The 'keyed bush' invented by Alfred Renshaw and Hans Renold Ltd.

1928 C. G. Renold appointed Chairman, R. O. Herford Deputy Chairman, and H. G. Jenkins Sales Director, of Hans Renold Ltd.

1930 Merger of the Renold and Coventry companies as the Renold and Coventry Chain Co. Ltd.; Chairman, Arthur Brampton (died 1955, aged 90); Deputy Chairman and Managing Director, C. G. Renold, (died 1967, aged 83); Deputy Managing Director, E. H. Kendrick (died 1949); General Works Manager, R. O. Herford (died 1955); General Sales Manager, H. G. Jenkins (died 1961).

1931 A pair of Holroyd wormgears achieves a world efficiency record at the National Physical Laboratory.
Renold and Coventry shares quoted on the London, Manchester, Birmingham and Glasgow Stock Exchanges.

1936 Holroyd shares granted quotations on the London and Manchester Stock Exchanges.

1937 Renold and Coventry sell the Brampton cycle fittings business, other than chain.
Mr C. T. Bayliss in charge of all Perry production.

1939 Outbreak of the Second World War.

1940 Hans Renold made D.Sc., *honoris causa*, of Manchester University.
Coventry and Manchester works bombed; Calais works under German control.

1941 R. O. Herford appointed Controller of Inspection (Administration), Ministry of Supply.
E. J. Parish (died 1968) appointed Acting General Works Manager.

1943 Death of Hans Renold, aged 90.
C. G. Renold becomes Chairman in succession to Arthur Brampton, who retires.
Holroyd acquire John Mountford and Co. Ltd., forgemasters, Manchester.

1944 R. O. Herford returns from Ministry of Supply to become Managing Director.

1945 The Perry Chain Co. Ltd. formed as a Perry subsidiary.
Formation of Perry and Co. (Holdings) Ltd.

1946 Holroyd acquire Manesty Machines Ltd., manufacturers of tablet-making and pharmaceutical machinery, Liverpool.
C. G. Renold appointed first Chairman of the British Institute of Management.
Mr W. S. C. Tully, CBE, appointed General Administration Manager of Renold and Coventry.
Five-day week introduced by Renold and Coventry.
Factory acquired at Cardiff by Renold and Coventry.

1948 C. G. Renold knighted.
H. G. Jenkins retires from office as General Sales Manager, and becomes Managing Director of a subsidiary.
Mr O. Bertoya appointed General Sales Manager and a Director.
Mr W. S. C. Tully appointed a Director.
Mr C. T. Bayliss and his brother, J. B. Bayliss (died 1956), Joint Managing Directors of Perry and Co. (Holdings) Ltd.

1950 Holroyd begin pump- and compressor-rotor milling, and the manufacture of milling machines for rotors.
First post-war Sales Conference of Renold and Coventry.

1951 Retirement from executive service of H. G. Jenkins.

1953 Retirement from executive service of R. O. Herford, OBE, and E. J. Parish.
Mr O. Bertoya appointed Managing Director.
Mr W. S. C. Tully appointed Deputy Managing Director.

1954 Opening of Renold House, Wythenshawe.
The company's name becomes Renold Chains Ltd.

1955 Mr W. V. Foley, General Sales Manager, and Mr L. J. Tolley, General Works Manager, appointed Directors of Renold Chains.
Renold Chains acquire the Anchor Chain Co. Ltd., Oldham.
Renold Chains acquire factory at Burton-upon-Trent.
Sir Charles Renold completes 50 years' service to the company.
Renold sales and manufacturing expansion in Australia.
Renold authorized share capital, £2,700,000; issued share capital, £2,693,072, consisting of £279,608 6 per cent cumulative preference stock (£1 units) and £2,413,464 ordinary stock (£1 units).

1956 Renold authorized share capital increased by £3,300,000 to £6,000,000.
To finance expansion of capacity, 1-for-2 rights issue of 1,206,732 ordinary of £1 each at 32s 6d. Issued share capital, £3,899,804.

1957 Agreement with the Formsprag Company, Michigan, USA, for Renold manufacture under licence of the sprag clutch.
Mr J. P. Halpin appointed a Director.

1958 Renold authorized share capital increased by £2,000,000 to £8,000,000.
To finance expansion of capacity, 3-for-8 rights issue of 1,379,804 ordinary of £1 each at 30s.
Issued share capital £5,279,608.
Offer made to all stockholders of Perry and Co. (Holdings) Ltd. to acquire their respective holdings. Acquisition in 1959 involved issue of £300,874 preference and £320,724 ordinary.
Renold begin wheel manufacture in Canada.
Holroyd begin the construction of new works at Milnrow.

1959 Issued share capital £5,901,206, consisting of £580,482 6 per cent preference stock (£1 units) and £5,320,724 ordinary stock (£1 units).
Mr C. T. Bayliss appointed a Director of Renold and Chairman of Perry.
Official opening of Renold Works, Burton-upon-Trent.
Mr W. V. Foley appointed Chairman of the Renold Australian subsidiaries.

1960 Renold authorized share capital increased by £2,000,000 to £10,000,000.
£2,660,362 of reserves capitalized and applied to scrip issue of one ordinary of £1 for every two ordinary held.
Issued share capital £8,561,568, consisting of £580,482 6 per cent preference stock (£1 units) and £7,981,086 ordinary stock (£1 units).
Mr R. Bate (General Sales Manager, Renold) appointed a Director.
Expansion, resiting and restyling begins of Renold branch premises in the United Kingdom.
The Perry pen manufacturing business sold.

1961 Sir Charles Renold appointed a Vice President of the Manchester College of Technology (now the University of Manchester Institute of Science and Technology). The new MCT lecture room block named 'the Renold Building'.

1962 Retirement from executive office of Mr O. Bertoya, Managing Director of Renold.
Mr L. J. Tolley appointed Managing Director in succession to Mr Bertoya.
Retirement from executive office of Mr W. S. C. Tully, Deputy Managing Director of Renold.
Mr J. P. Halpin appointed Deputy Managing Director and Finance and Overseas Subsidiaries Director of Renold.
Mr E. E. Hempsall, General Works Manager, appointed a Director of Renold.
Renold acquire an interest in local manufacture, Spain, following an earlier licensing agreement.
New Holroyd works at Milnrow in operation; extensions begin at Holfos Works, Rochdale.

1963 Renold acquire Arnold and Stolzenberg GmbH.
Renold product manufacture in UK rationalized.
Willingness to join a trade union accepted as a condition of employment in Renold UK works establishments.
The company establishes the Sir Charles Renold Fellowship at the Manchester College of Science and Technology to mark Sir Charles Renold's eightieth birthday.
Renold acquire an interest in local manufacture, Argentina.
Renold selling subsidiary established in Austria.

1964 Renold authorized share capital increased by £3,000,000 to £13,000,000.
£2,660,362 of reserves capitalized and applied to scrip issue of one ordinary of £1 for every three ordinary held.
Issued share capital £11,221,930, consisting of £580,482 6 per cent preference stock (£1 units) and £10,641,448 ordinary stock (£1 units).
Renold authorized share capital increased by £4,000,000 to £17,000,000.
First Renold Regional Sales Conference, Houston, Texas.
Renold acquire all the share capital of John Holroyd

and Co. Ltd. for £6,025,000 in cash and £3,000,000 Renold ordinary stock.
Mr W. V. Foley returns from Australia and rejoins the Renold parent Board.
Creation of the Renold Executive Manager Grade.
Acquisition and installation of an English Electric-Leo-Marconi Leo III computer at Renold House.

1965 Mr E. O. Liebert and Mr G. W. Taylor, Directors of Holroyd, appointed Directors of Renold.
Issue of £3,000,000 of $6\frac{3}{4}$ per cent debenture stock, 1990–1995.
Issued share capital £14,221,930, consisting of £580,482 6 per cent preference stock (£1 units) and £13,641,448 ordinary stock (£1 units).

1966 Decision to build new Renold works at Thornaby-on-Tees.
Issue of £3,000,000 of 8 per cent debenture stock, 1991-96.

1967 The new Group organization in being; the company takes the name of 'Renold Ltd.'
The Queen's Award to Industry conferred upon Renold, for export performance, and Holroyd, for technological innovation.
Foundation stone of the new Research and Development buildings at Renold House laid by Sir Charles Renold, May 4th.
Sir Charles Renold retires from the chairmanship of Renold Ltd. on August 3rd, and becomes Honorary President of the company, remaining a Director.
Sir Thomas Robson, MBE, appointed Chairman.
Renold acquires Crofts Engineers (Holdings) Ltd.
Death of Sir Charles Renold (aged 83), on September 7.

The following are the present (1968) Directors of Renold Ltd.:

Sir Thomas Buston Robson, MBE, Chairman.
Leslie John Tolley, Group Managing Director.
John Philip Halpin, Deputy Group Managing Director and Group Finance Director.
Rex Bate, Renold Sales Divisional Managing Director.
Charles Arthur Percival.
Edwin Garlick.

Oscar Bertoya.
John Arthur Croft.
Sir Patrick George Hamilton, Bt.
Edward Eagle Hempsall.
Sir Leslie Harold Robinson, KBE, CB.
Granville Whittaker Taylor.
William Scarth Carlisle Tully, CBE.

CHAPTER I

RENOLD RETROSPECT, 1879–1955

The history of Renold Chains Ltd., from its origins in the work of Hans Renold until the year 1955, was described in some detail in a book published twelve years ago.[1] The narrative told of the founder's entry at an early age into the business of making chain drives for the Lancashire cotton industry, his invention and patenting of the bush roller chain, and the extension of his small business into wider fields of precision chain manufacture and application for many kinds of industrial drives. Hans Renold's emergence as a leader in specialized power transmission, his inventiveness as a mechanical engineer, and his progressive and humane outlook as an employer, were set out. The merger in 1930 of the company he had formed in 1903, Hans Renold Ltd., with that of his principal competitor, the Coventry Chain Co. Ltd. (which had itself acquired a third strong contestant in the chain business, Brampton Brothers Ltd. of Birmingham, in 1925) was explained. The account covered the fortunes of all three companies in peace and war, and the evolution, with the endorsement of Hans Renold (whose influence remained strong until he was nearly ninety, at which age he died in 1943) and under the leadership of his son, the late Sir Charles Garonne Renold,[2] of a philosophy of direction, management and technology. All this spanned the years, in terms of precision chain manufacture, from 1879 to 1955. Yet 1955 marked no finality; in the second half of the century, between 1956 and 1967, momentous events have occurred in the history of Renold, presenting many aspects of interest and

[1] Basil H. Tripp: *Renold Chains; A History of the Company and the Rise of the Precision Chain Industry, 1879–1955*, London, Allen and Unwin, 1956.

[2] Charles Renold was knighted in 1948.

significance. A record of them may serve to show how one world-wide, specialized technical product company has found a way of life during an epoch of unparalleled rapidity in industrial and technological development.

This is the reason for a second book devoted to the present-day Renold Group and its fortunes. But first it is necessary to look back over the earlier and longer period, to obtain a picture in outline of the circumstances confronting those responsible for guiding the company from the late 1950s to the present time.

In 1954 the company's offices at its principal establishment, Renold Works, Didsbury, a suburb of Manchester, were transferred to an entirely new building, Renold House at Wythenshawe, about ten miles from the city centre. This was a stage in a process that had begun when Hans Renold, a frugal and hard-working Swiss and a mechanical engineer of genius, after some years in his native land and in France, came to England and in 1880 invented and patented the bush roller chain. The invention was crucial because it provided the foundation for development in driving chain techniques applicable in manufacturing industry and transport.

Two features were necessary in the design of a successful driving chain, namely, efficient articulation in the chain itself, and good gearing with the wheels. Their achievement presented similar engineering problems, whether the chain was needed for high- or low-speed power transmission; the difficulty of meeting both needs became more acute as the economic advantages of fast running came to be appreciated. A maker of chains for textile machinery drives used in the Lancashire cotton industry, James Slater, was the first patentee of a driving chain in 1864. He went some way towards solving the wheel-gearing problem with what was known as his 'bowl' chain (the 'bowl' being a free roller on the transverse pins of the chain), and efforts were made by other inventors. Hans Renold, who bought Slater's business in 1879, improved upon the Slater invention or, to put it more accurately, thought out the problem anew. Instead of permitting the connecting plates between the pins to articulate directly with them, each pair of inner plates was mounted on the ends of bushes. This meant that articulation occurred between the pin and the bush, the bearing surface of the roller in which the bush worked extending across nearly the whole width of the chain. The design met both the necessary features mentioned above—efficient articulation in the chain together with good gearing. It gave scope also for embodying improvements in metallurgy and manufacturing methods, and remains fundamental to current precision chain design, manufacture and use.[1]

A shrewd judge of men, Renold gathered round him more than one generation of competent lieutenants; he had also the good fortune to be the father of Charles Renold, who, with some of the paternal aptitude for engineering, was certainly the older man's equal, if not his superior, in business and affairs. The product of these factors was decisive after 1903, when Hans Renold formed his firm into a company with the name of Hans Renold Ltd., which soon came to stand higher than its main British competitors alike in the quality of its products and the inventiveness of its engineers. To Hans Renold the professional standards of an engineer and the craft standards of a skilled workman were matters of the utmost importance. Throughout a long life he declined either to permit, or to turn a blind eye to, the manufacture of an inferior article at a cut price, refused to sell chains or chain drives to any would-be customer whom he thought might misapply them, and insisted upon his right and, indeed, duty as the manufacturer to advise upon their correct employment. This doctrine, sometimes carried to the point at which much-needed business was lost, is taught and practised over a wider range of products in the Renold Group today.

For these reasons and others, arising from technological and economic development at large, Hans Renold Ltd. grew steadily in peace, and passed successfully and with credit through two great wars. By degrees there were incorporated with it the two chief competitors in England. Either separately or together, these had on the whole fought a losing battle with Renold in the severe rivalry for the growing market. How this

[1] See also Chapters 2 and 3, *Renold Chains* (op. cit., p. 25).

came about has been told in some detail in the book previously cited, but it must be mentioned here as the preface to a summary of the situation at the close of 1955, the year before the phase of Renold history with which the present volume is concerned.

In 1925 there took place the first big merger of the two principal rivals to Renold—Brampton Brothers Ltd. of Birmingham, with their French subsidiary, SA des Etablissements Brampton, Calais, and the Coventry Chain Co. Ltd., of Coventry. Brampton Brothers had entered the cycle industry about 1885, as cycle fitting and chain manufacturers, chain being part only of the business. The Coventry company had been formed in 1896, successfully staking a claim to a share of the cycle chain and freewheel market. Each had given wide scope to the talents of able leaders, the chief of whom were Arthur Brampton in the one case and Alick Hill in the other. Hill, who was Mayor of Coventry in 1916, died in 1921. Arthur Brampton lived much longer, dying in 1955 aged 90, twelve years after the death of Hans Renold at the same age.

During the years between the entry of these two companies into the driving chain market and their merger in 1925, however, Hans Renold Ltd. had not stood still. In the works, and in the markets of the world, the company had shown itself to be a first-class instrument of what may be described as rational perfectionism in its field. It felt the competitive struggle, its adherence to high standards sometimes caused it to be worsted, yet it retained the leading place in the precision chain industry. The Renold leaders were better at discerning the changing and developing trends of demand. They had a keen appreciation of the way in which successive extensions of mechanization—for example, the increasing use of oil and petrol engines for purposes other than road transport—were everywhere enlarging the scope for precision chain. They were never successfully tempted to depart from single-minded concentration on precision chain alone, the only non-chain additions to their product range being in reality the logical provision of ancillary equipment to make chain drives more effective. Product diversification was to come much later in greatly changed circumstances.

The deliberate dedication of the Renold effort led steadily to the successful discovery and advocacy of new chain applications. This resulted in continual expansion of production capacity, with the aid of machines often tooled and engineered from within the company's own technical resources and owing much to the creative ability of Hans Renold himself. Some of his designs contributed substantially to general machine-tool development in the United Kingdom. Profits were consistently ploughed back; for many years no appeal was made for publicly-subscribed capital. The company kept abreast of the developing markets as chain driving extended from the cycle to the car and motor-cycle, moved into industrial driving, and began to be adopted for conveying and mechanical handling. Hans Renold Ltd. knew their markets and their customers, seized the advantages offered by mass production and standardization and clinched them with the maintenance at tactical points of adequate supplies of 'stock drives'—standardized chains and wheels capable of meeting a variety of industrial norms. Over all these by then multifarious activities the founder's principles were firmly applied: the insistence upon high quality, the determination to educate the customer in sound application.

It was well before the First World War, in 1909, that Hans Renold Ltd. founded a 'Social Union' to encourage friendly feeling among the workpeople of all grades. The official business of engaging staff and assisting them with their personal problems of adjustment and well-being, later to become commonplace in big companies, but then comparatively rare, was begun with the creation of a personnel department in 1910. The war in 1914, although it meant a full-scale transformation of the company into a great complex for munitions production, as did the Second World War in 1939, brought no halt in the march towards fulfilment of long-cherished ideals in the sphere of rational and civilized labour relations. Before the 1914–18 War, Renold working hours were shorter than those generally ruling in the engineering industry; profit sharing was firmly established by 1922. The conception that capital and labour had common tasks and interests had already received a stimulus in 1917, when the policy of regular joint consultation between management and the labour force, the latter represented by the accredited shop stewards of the trade unions concerned, came into being and flourished, after its initial difficulties, under the care of Sir Charles Renold, who made it his special study.[1] Amenities in health and sickness, welfare and recreation, kept pace with this directorial and managerial outlook upon the company's responsibilities, clearly illustrating a dictum of Hans Renold: 'Our job is not to make chains; it is to make men and women—they will make the chains for us.' Be that as it might, he did not neglect any technical assistance that they might need in the performance of this task.

'Constitutional management' is now built into the Renold system in Britain, and with the modifications made necessary by local legislation applies in other countries where the Renold Group makes or markets its goods by means of its own factories, warehouses and staffs. It successfully stood the test of the Second World War and the difficult conditions thereafter, and in 1968 is as vigorous as ever. The result is an immensely strong, yet sensitive and flexible instrument.

The Coventry-Brampton merger—in financial terms the acquisition by Coventry of the Brampton share capital—was partly a consequence of the competitive power of Renold. But the product range of the three companies overlapped at many points, and rationalization in the interest of quantity production, price and efficiency was clearly necessary. Consequently, the Coventry-Brampton merger proved to be a step on the road to a junction of resources of what were, for all practical purposes, the three leading chain manufacturers in the United Kingdom. In many respects, with variations in policies affecting prices and quality, all three companies were interested in the same kind of product. The advantages to be gained within the companies and in the national interest by combining their resources had been apparent to Sir Charles Renold as early as 1919. He worked steadily for ten years to bring about the unification, in the later stages with the added

[1] See Sir Charles Renold: *Joint Consultation over Thirty Years; A Case Study*, London, Allen and Unwin, 1950.

power derived from the Chairmanship of Hans Renold Ltd., in which office he had succeeded his father in 1928. Two years later the merger was completed; Coventry, already including Brampton, and Renold, became a single industrial and commercial entity.

To outward seeming, this achievement may have worn an air of inevitability, but in reality it was the outcome of protracted negotiations periodically faced by many difficulties. Once it had been completed, the directors of the Renold and Coventry Chain Co. Ltd., as the new company was named, faced the task of channelling the different streams of tradition and bringing about a smooth confluence of staffs and workpeople. Arthur Brampton was appointed to the chairmanship of the new organization. The successful achievement of unity was in considerable measure due to his broad views and genial spirit.

The advantages claimed for a unified organization were fully realized, whether they touched the interests of the shareholders, the efficiency of the various factories, the improvement of the product, or the requirements of the customers in the United Kingdom and abroad. The driving chain and its ancillary equipment became the single product of an avowedly specialized company concerned with a particular aspect of power transmission. Little by little what had become the sidelines were disposed of in one way or another—for example, the Brampton cycle fittings interests were sold to another company and the Coventry freewheel business was abandoned. The concentration on chain and chain driving was to be maintained until it joined logically with another form of power transmission more than a generation later, as we shall see.

By 1955, with the general and sales direction and administration operating centrally, in daily touch with all the movements and actions of a manufacturing business extending throughout the world, Renold Chains Ltd. was employing a total of over 6,000 people in its British works in Manchester, Coventry and Cardiff, at those in France and Australia, in its selling subsidiaries in Belgium, France, Germany and Holland,

and in Canada. It had created a network of selling agencies in many overseas territories, commonwealth and foreign; in the same year a new factory was opened at Burton-upon-Trent. An effective board structure had been evolved, directing a business with a single doctrine governing the quality of its products, its pricing policy, and the obligations it felt to all whose lives it touched at any point. The elements constituting the board consisted of men with wide industrial, financial and administrative experience from outside, of retired but active seniors who had made their careers inside the organization, and of the executive heads, chief among whom by 1956 were the Managing Director, Mr Oscar Bertoya, and the deputy Managing Director, Mr W. S. C. Tully. Mr Bertoya had served the company since 1923, principally in the sales division; Mr Tully, after a business career in India, had been appointed in 1946 as General Administration Manager, and finally to the deputy Managing Directorship. Over them presided Sir Charles Renold, who had been a director for close on half a century and who possessed longer experience than any of his colleagues.

It had now become established practice that daily direction and supervision should be the duty of a compact and balanced team at the summit of executive responsibility. Sir Charles Renold, with Mr Bertoya, Mr Tully, Mr W. V. Foley, the General Sales Manager, later succeeded by Mr Rex Bate, Mr L. J. Tolley, who had joined Renold Chains after achieving a high reputation with the Nuffield Organization, and Mr J. P. Halpin, who succeeded Mr Tully as General Administration Manager in 1956, were the leaders who took the company into the 1960s.

Financially, the year that ended on April 3, 1955, had produced a trading profit of over £1,000,000, an increase of £417,800 compared with its predecessor. Demand had been heavy, sales were a record and had been matched by output, but the point had been passed at which cost increases could be offset by greater productivity. Works extensions planned in 1951 had shown their worth, and modernization and re-equipment were continuing. Nevertheless, the rise in costs

HANS RENOLD

and the Government's taxation policies cast shadows over the scene.

In the period since the modest beginnings of 1879 the Renold community has grown steadily in scale, statistically and operationally. It has met and overcome many of the problems common to developing communities. It has now attained a position, within its strictly-defined terms, as a world-wide technical product group of companies concerned with certain aspects of power transmission. With the passage of time, the Group has become an integral part of international industry. For this reason, as also because of the successful techniques it has evolved to fulfil the arduous requirements of its position, it is worth studying. The premise holds for the years 1956–67, the subject of the present book, no less than for those spanning the period between Hans Renold's first small venture into private manufacturing and the opening of Renold House in 1954 as a self-contained administrative headquarters for the company.

The role outlined above was extended and enhanced by the original Renold company's emergence as an international positive power transmission group, a role made more important by mergers with three other large companies, Perry and Co. (Holdings) Ltd. in 1958, John Holroyd and Co. Ltd. in 1964, and Crofts Engineers (Holdings) Ltd. in 1967. The last-named company, however, falls outside the period with which these pages are concerned. With Perry there came into the Renold fold, as had happened with the Coventry Chain Co. Ltd., another powerful precision chain manufacturing organization. With Holroyd, Renold crossed the dividing line between chain transmission and speed-reduction gearing, thus bringing about a technical alliance between two complementary drives, and uniting the marketing interests of two dissimilar organizations that had nevertheless long worked together as a result of the positive drive affinity between chain and wormgear.

CHAPTER 2

THE PERRY CONTRIBUTION

Readers of the earlier Renold history will recall that Isaac Brampton and his son Frederick (the real founder of the Brampton fortunes) went to Birmingham and began to make metal fittings for binding account books, among other useful accessories of one kind and another, until Isaac's grandsons entered the young and expanding cycle industry, thus being prompted to take up the manufacture of chain. It is a curious coincidence that Birmingham and the stationery trade provided the first openings for the Perry companies also. Certainly the latter had a history as varied and fascinating as any in British industrial annals.

Perry and Co. Ltd. was established in 1876, as the result of an amalgamation of three other firms. These were A. Sommerville and Co., of Lancaster Street Works, Birmingham, makers of steel pens (i.e. pen nibs), Wiley and Son, whose factory was in Graham Street, Birmingham, and who made pen and pencil cases or holders, and Perry and Co., of Red Lion Square and Cheapside, London, who were pen, pencil, elastic band and ink makers and merchants, and dealers in stationers' sundries.

The proprietor and in a decisive personal sense the dominant figure in A. Sommerville and Co. was by then Sir Josiah Mason, who had begun to make pens in 1828, long before he joined Sommerville and Co., whose business he eventually acquired. There is, or was, considerable controversy as to the original inventor of the pen or nib as we know it, with its present-day variations; if the subject were to be explored here in detail we should speedily be lost in antiquity and later find ourselves in a maze of conflicting opinions about the primacy among pen makers in the eighteenth and early nineteenth centuries. It is reasonably certain, however, that Sir Josiah Mason was the first manufacturer on a large scale of the inexpensive steel nib, for which he deserved the thanks of a host of those formerly and accurately described as quill drivers, whose working days were burdened with the problem of cutting goose-quills accurately to a writing point that would last for a few pages of a letter, a ledger or a manuscript. Sir Josiah's pens and penholders were sold through Perry and Co. in London, whose own pens had long been marketed under the trade name of the 'Perryian' pen, and were already well known. The original Perryian pen has acquired a modest fame in *The Ingoldsby Legends*, first published in 1840, in which the Rev. Richard Barham (whose robust and humorous verse, written under the name of 'Thomas Ingoldsby', is still popular) refers to a contemporary theatrical controversy, turning on the rival merits of two singers and involving an impresario named Doldrum. He records how:

> 'Dol-drum the Manager, full of care,
> With a troubled brow and dissatisfied air,
> Saddest of men, sat down and then
> Took from his table a Perryan pen'

—and wrote a letter about his troubles to a newspaper. Despite the omission of the penultimate vowel 'i' in 'Perryian' (the correct trade spelling) there can be no doubt about the factory whence came the pen that Mr Doldrum took up.

The Perry company's association with Mason began because the former found that Mason's pens sold better than their own. Perry's therefore gave up pen manufacture, and their contracts with Mason provided that he should be their sole supplier, but that they should continue to be called Perry or 'Perryian' pens. A somewhat similar arrangement for the supply of pen and pencil cases linked W. E. Wiley, of Wiley and Son, with Perry. The heads of the Perry business were James Perry and his brother Stephen, who were born at Banbury, Oxfordshire, at the turn of the eighteenth century. Sir Josiah Mason had been rich enough in 1870 to buy the Sommerville business upon the retirement of the proprietor, Alfred Sommerville, and at the time of the amalgamation in 1876 he sold it to the newly formed Perry and Co. Ltd. for the considerable sum, in golden sovereigns, of £130,000, the Wileys receiving £36,000 for their business. Stephen Perry's two sons, Joseph and Lewis (to whom the Perry company proper had come after the death of their father in 1873), obtained £80,000 for theirs. Accompanying the amalgamation were agreements as trustees with Joseph Henry Nettlefold, screw manufacturer, thus linking the Perry company with another famous Birmingham name, and Francis Adkins, of the Brades Steel Works, now part of a group of edge tool manufacturers. The shares allotted subsequently to the Perry amalgamation went mainly to customers, friends of the principals, and home and foreign agents and few were publicly subscribed.

Sir Josiah Mason, according to the amalgamation prospectus, gave as his reasons for the sale of his interests, his old age and his wish to retire after fifty-three years in business; he died four years later, on June 16, 1881. With him there departed both a benefactor of Birmingham and a great man of business cast in the rugged and individualistic mould whence came so many leading industrialists of his time. Among his benefactions was the endowment of Mason College, from which grew the University of Birmingham. As a manufacturer he was notable not only for pens, but in the field of electroplating and gilding, in which he was associated with G. R. Elkington, who made the process a commercial success. The later development of the pen manufacturing business, however, owed much to his nephew, Isaac Smith.

The manufacture of pens demanded skilful mechanical working and heat treatment of the steel sheet which was the basic raw material, and accurate presses to stamp out the blanks and slit them before grinding. In these processes originated the engineering tradition of the works, which was fostered by Mason's successors. The engineering skill, of course, resided principally in the toolmaking required and the maintenance of the machinery, including three steam engines driving the line shafting. The workforce employed in the preparation of the presses consisted principally of women; it is said that the company was attacked by the romantic novelist Marie Corelli, who criticized the conditions under which they worked and the payment they received. She had apparently expressed a wish to see the works, and had been welcomed by the management with what would nowadays be called 'V.I.P. treatment'. A new pen was named after her; it was produced in quantity, in readiness for what were expected to be her encomiums. The business instinct to make the most of her visit was sound enough, but when the result appeared in print it was far from being a good advertisement. The 'Marie Corelli' pen therefore ended as works scrap.

The fact seems to be that by the standards of the age the conditions were by no means disgraceful. For example, there was a works canteen with tables for 600 people, where a hot meal was served at prices ranging from 2d to 6d. From a stage at one end of the room dramatic entertainments and concerts were given during the winter months by amateur performers among the operatives. At the other end was a library containing 2,000 books issued free on loan to employees, of whom there were about 1,000 producing 58,000 gross of pens weekly. No records are available of the hours worked nor the wages paid, but the foregoing description of the amenities falls short of the 'dark Satanic mills'. In the 1890s there were thirteen pen and metallic penholder manufacturing firms in Birmingham, and a contemporary account states that in the trade generally women earned from 4s to 15s weekly, boys from 5s to 10s, unskilled men from 18s to 30s, and toolmakers from 25s to £3. Most of the women were on piecework, but the men were on time rates. These are low wages indeed by our standards, but in the setting of their day they compared not unfavourably with others.

After the 1876 amalgamation of the Sommerville, Wiley and Perry companies as Perry and Co. Ltd., the Wiley pencil-case works was moved to Lancaster Street. A strong card in the Wiley hand was the patent taken out by W. E. Wiley in 1857 for what would now be called a propelling pencil. One of the materials used for the case containing the lead was 'aluminium gold', a composition (states a booklet issued by Perry and Co. in 1892) 'so resembling gold that it cannot be distinguished from it'—which brings to mind the Bishop who advocated a teetotal beverage with the memorable comment that 'it looks like beer, it smells like beer, and it tastes like beer, but it is not beer'.

During the 1890s Perry and Co. Ltd., under its first Managing Director, Joseph Perry mentioned above, the son of Stephen, had integrated its manufacturing work at Lancaster Street, Birmingham, and its stocks and distribution arrangements in new offices and warehouses at Holborn Viaduct, London. The company continued to have a strong interest in the stationers' sundries side of its trade, and in the successful manufacture and marketing of Perry pens. But in the meantime the safety-bicycle had been invented and the Perry Board were not long in deciding to try their fortunes with the cycle manufacturers. They began to produce chains, hubs and freewheels. The leading part in successfully adding these new lines to the traditional range was taken by J. W. Bayliss, who for many years managed the whole of the cycle accessories branch of the business; in 1926 he became a director, and died in his eighty-fourth year, in 1951. Bayliss made Perry chains a competitive factor of importance.

Although in the early years of the cycling boom the company produced an extensive range of cycle components and accessories including chains, hubs, frames, forks, brackets and lugs, chain wheels and cranks, brake fittings, pedals and the like, it was to the manufacture of chains that their resources were principally directed and this soon became their main product. Perry first produced a cycle chain in 1884 and this was followed in 1889 by the manufacture of a 'block' chain, in which the inter-link was a block pierced to form the pin bearings. This incorporated a patented bush produced from pen steel, designed to prevent the rivets turning in the side plates of the chain and to transfer friction from the soft steel rivet to a hardened steel bearing. A year later, the company began making roller chain. In 1894 Hans Renold's patent for the bush roller chain expired, and the advantages of the invention

MR C. T. BAYLISS

became available to others. Renold had supplied cycle chain in 1-in. pitch, believing at the time that the smaller size of 0·5-in. pitch, in which Bayliss had faith, could not be manufactured in a way that would give satisfactory service. Bayliss proved that a 0·5-in. pitch cycle chain was suitable, and in 1899 the Renold company also began to make it.

The growth in demand led to the erection of a separate factory for cycle products on land adjacent to the main works. This sufficed from 1898 until 1907, when, to meet continuing expansion, Perry acquired an eleven-acre site at Tyseley in the Birmingham suburbs, on which was built a still larger establishment. This became the chief manufacturing centre for the company's cycle and motor-cycle products.

In 1901 Perry and Co. introduced a ball-bearing cycle freewheel to replace the previous fixed-wheel sprockets. Eight years later they began making coaster hub brakes, much favoured by cyclists abroad, especially in North America. A factory in premises leased from the Board of Trade at Abercrave, South Wales, was used for this purpose many years afterwards.

During these early years of the twentieth century, a remarkable young engineer joined the Perry staff as a machine designer, Mr C. T. Bayliss. He was born in 1892, the younger son of J. W. Bayliss, and received his training in the drawing office of H. W. Ward and Co. Ltd., machine tool manufacturers, Birmingham. When he came to Perry in 1910, he was the only draughtsman employed by this large company with its diverse manufacturing interests and production problems; his father, and the other senior engineers, were their own draughtsmen and designers, in a way that would be impossible today.

Mr Bayliss had plenty of ideas which he soon began to drive home at Perry's. He possessed for his age a rare insight into what machine tools might be made to do in the era of mass and batch production then dawning. An early convert to the internal combustion engine as a prime mover, he saw clearly the engineering changes it was bringing about. His belief in the motor-car coupled with his own experience as a motor-cyclist and light car user and driver led him to design what soon became known as the Perry car and caused Perry to add automobile manufacture to their heterogeneous list. Mr Bayliss was still in his teens when he designed the first of these vehicles; by the autumn of 1912, when he was twenty, the Perry Motor Co. Ltd. had been registered, and early in 1913 was producing cars at Tyseley. The capital of the company was £60,000, of which Perry and Co. Ltd. had invested £50,000. The first Perry car was a 6 h.p. twin-cylinder model selling at £137.[1]

At this time the Perry Chairman, Martyn Smith, a grandnephew of Sir Josiah Mason, was pleased enough with Mr Bayliss's design to present him with a gold watch inscribed in commemoration of his achievement and recording the fact that he was only in his twenty-first year.

The car was popular. It sold well, and its reputation soon became known to William Morris, the future Lord Nuffield, who not long before had begun his own career as a motor manufacturer. At the 1913 Motor Show, Morris sought a meeting with the designer of this new challenger in the light car market.

'My God, you are a kid, aren't you?' said Morris when they were introduced. 'How many cars did you make last year?'

'Three hundred,' replied Mr Bayliss.

'What!' said Morris, 'I only made two hundred.'

It is tempting to speculate on the effect had Perry and Co. backed Mr Bayliss to the fullest extent possible, thrusting decisively into the young but fast-growing motor industry. But, despite the fact that Mr Bayliss had been appointed works manager and chief designer of the Perry Motor Co. Ltd., and had laid out the Tyseley motor-car factory for production, including manufacture of the complete engine, gearbox, universal joints, front and rear axles and steering unit, everything, in fact, except the tyres, magneto and chassis, there was a fundamental divergence between his own views on motor-car production and those of Martyn Smith. Mr Bayliss had realized that production of the car might be simplified, made more

[1] For an account of the technical features of the Perry car, see Lytton P. Jarman and Robin Barraclough, *The Bullnose Morris*, London, Macdonald, 1965, page 76.

economical and therefore more profitable if certain standard engine components—valves, for example—could be bought in bulk from outside manufacturers better able to produce them in the quantities he already saw to be necessary, and assembled at Perry works. The economic basis for making the complete engine under one roof, upon which Martyn Smith insisted, was unsound, for if it were to be successful, it called for greater resources in plant and equipment than the parent company either could or would provide. Mr Bayliss's arguments for calling in outside suppliers, thus making it possible to produce more cars more cheaply, and in no way inferior to the nearly 100 per cent Perry product, gained no support; they were, indeed, rejected roughly. The scales were weighted against him also by the outbreak of war in 1914. An 11 h.p. four-cylinder Perry was introduced in 1915, but by 1916 it cost £260. This was £100 more than its nearest rival from the Morris stable, the Cowley, so that Morris possessed the competitive edge over the 'kid' who had surprised him in 1913.

Moreover, by this time Mr Bayliss was involved in munitions production. When the war broke out he had grasped at the opportunity, as he thought, of fulfilling the twin ambitions of joining up and learning to fly. He therefore applied for a commission in the Royal Naval Air Service, later to be merged with the Royal Flying Corps as the Royal Air Force in April 1918. He was in due course called before a board of interviewing officers, and was gratified to see a Perry car outside the door of the building in which the interview was to take place. The presiding officer asked him what he did in civilian life.

'You say here,' he said, pointing to the letter of application, 'that you are an engineer; what does that mean? What sort of engineer?'

'I am a motor-car designer,' replied Bayliss.

This was received with incredulity; he was tartly informed that one so young could not possibly have designed a motor-car; in any case, if he had, which motor-car?

'There's one outside the front door now,' said Bayliss.

'What, the Perry car?' asked the president. 'That's mine. A very good car too. But I don't believe it. Can you prove it?'

In answer, Mr Bayliss produced the inscribed gold watch. This carried conviction. It also meant the end of his hope to be a pilot; the interviewing board, and through them the gods in Whitehall, decided that engineering brains of that order could be turned to better account.

After the 1918 Armistice and because of the disagreement on engineering policy between the Perry Board and Mr Bayliss, the latter joined forces with a friend, Arthur Edward (nicknamed 'Archie') Wiley. He was a descendant of the Wileys of propelling pencil fame, and a man of considerable business talent—which he subsequently used with success in the brewing trade until his death in 1964 at the age of seventy-six. In 1919 Mr Bayliss and Archie Wiley purchased all the assets of the Perry Motor Co. Ltd. except the name. By that time, although about 1,000 Perry cars had been manufactured, Martyn Smith's reaction was on the whole one of good riddance to a branch of the business beyond the scope of his preconceptions. But soon afterwards the two young men sold their newly acquired interest to A. Harper, Sons and Bean Ltd., who paid a handsome price, and placed the 11-horse Perry on the market later in the year, when it became the popular Bean car well known in its day.

Mr Bayliss and Wiley then set up as manufacturers of cycle bearing components under the name of Bayliss, Wiley and Co. Ltd., with fair success.

In 1926, Mr Bayliss and Wiley returned to the Perry fold, in which their company was merged. Mr Bayliss remained in charge of the Bayliss, Wiley works until his appointment as a director of Perry and Co. Ltd., in 1937, when he became responsible for all the company's production and for the design of special-purpose machines in the chain and cycle components branch of the Perry Group. During the Second World War, when munitions production was again the first call on the company, one of their chief contributions was the manufacture of 2-lb. pom-pom shells for the Royal Navy.

In 1945, the company transferred the chain-making and coaster hub business to a newly formed subsidiary, the Perry Chain Co. Ltd., the financial control of the entire Perry Group then vesting in a parent company named Perry and Co. (Holdings) Ltd., from 1948 under the chairmanship of Wiley, Mr C. T. Bayliss being joint Managing Director with his brother, James Bell Bayliss, who died in 1956. The field was becoming diverse and much extended: the chain company made chain for cycles, motor-cycles, camshaft drives, and for industrial and agricultural driving; there was the coaster hub factory at Abercrave, and the firm of Bayliss, Wiley, in its role as a subsidiary, manufacturing cycle fittings. Moreover, the holding company controlled also the totally dissimilar business of Perry and Co. (Pens) Ltd. There were in addition three overseas subsidiaries to be nurtured. Perry Bayliss Wiley (Belgium) SA, Brussels, was formed in 1946; Perry Bayliss Wiley (USA) Inc., New York, and Perry Bayliss Wiley (Canada) Ltd., Ontario, were added in 1948. These overseas selling and distributing companies were set up because something like 50 per cent of the Perry production was exported. The New York undertaking was also, in part, a response to the United Kingdom's shortage of dollar currency, and the British Government's insistence that industry should make exceptional efforts to increase the supply.

On the face of it, all this activity suggested that here might be the beginning of another of the large and powerful groups of federated interests, with a single financial control but a wide and differing range of products, nowadays a marked feature of industry and marketing. There was confidence and determination among the leaders, and a solid foundation of past successes upon which to build. What, then, in the ten years between 1948 and 1958, led to a readiness to merge with the larger chain manufacturing competitor in the north?

In the first place, with the advent of the fountain pen and later the ball-point pen, the nib business had been declining for years. The pen specialities with which Perry attempted to retain a foothold in the stationery trade proved insufficient to provide a substantial replacement.

Then practically the whole of the remaining Perry eggs were in the cycle basket. This proved a most bountiful one in the years immediately following the end of the Second World War, but after that cycle production fell away rapidly, cheap foreign components began to be fitted by British cycle manufacturers on the one hand, or were supplied by the makers' own works on the other. Practically the whole of the coaster hub production went to export, mainly to the United States of America, as home consumption was small. The USA outlet began to shrink as well.

On the chain side, Perry's stake was also predominantly cycle-oriented. They had a significant cycle chain production, but little else, and although they most shrewdly entered the moped trade, it did not provide a lasting way out of their difficulties. Consequently, unless they changed their business entirely, their only alternative was to widen their chain interests into a full range. This would not only have been expensive (although Perry in fact entered the motor-cycle and camshaft chain business), but there was no capacity shortage in the market, as Renold had laid their plans to match all foreseeable demand from British industry, including of course the motor industry.

In the circumstances, the sensible policy was to join forces with Renold, the more so as the sizeable Perry cycle chain production, coupled with Renold capacity, provided much scope for rationalization.

Thus matters stood when in 1958 Renold Chains Ltd. offered to purchase the share capital of Perry and Co. (Holdings) Ltd., and in 1959 the acquisition was completed. As in every previous merger in which the Renold company had been concerned, the ensuing rationalization of the former Perry activities was carried out with a careful regard for individuals. A tidying-up process began among the overseas subsidiaries; since Renold were widely and effectively represented in Belgium and in Europe generally, the Perry Belgian interests were terminated. The American subsidiary, however, undertook a new lease of life; in 1964 its name was changed to Renold Perry Inc., and in the following year there began its move to new premises on an industrial estate at Farmingdale, Long Island, NY, where there are increased storage capacity and better transport facilities. As was the case in Belgium, Renold representation in Canada made it unnecessary to continue the former Perry subsidiary in Ontario, and this was closed.

The main part of the Perry enterprise was assembled in 1959 into a single new company named Perry Engineering Ltd. The transition from the former ownership was made the smoother and happier by the appointment of Mr C. T. Bayliss as its Chairman, with a seat on the Renold Board. Apart from Mr Bayliss's eminence as an engineer and designer, he was trusted and popular among all the Perry staff and workpeople. His whole-hearted support of the new arrangements therefore gave them confidence and encouragement in what to many seemed a disconcerting development. In 1965 Mr Bayliss retired as Chairman of Perry Engineering and in 1966 from his Renold directorship. In his annual statement to the Renold shareholders in 1966, Sir Charles Renold spoke appreciatively of Mr Bayliss's wise counsel and the readiness with which he had co-operated in bringing the sound and in some ways outstanding Perry engineering tradition into the wider life of the Renold Group. Elsewhere, in an appreciation privately circulated within the Group, Sir Charles reminded his readers that Mr Bayliss had been 'a designer of some very striking machine tools, particularly for the cold forming of cycle bottom bracket axles and for the automatic assembly of chain', and referred to him as 'a friend and colleague—a shrewd and pleasant counsellor, a distinguished engineer and a staunch collaborator in the prosecution of Group policies'.

In the spring of 1961, Perry and Co. (Pens) Ltd. was sold to British Pens Ltd. It was clearly an oddity in the Renold field of positive power transmission.

At the end of the financial year in which pens ceased to concern the Group, the newly acquired Perry works had made a satisfactory contribution to Group results through its chain and cycle fittings production and sales, but the coaster hubs were causing anxiety, largely as a result of a recession in the United States, hitherto one of their chief outlets. In December 1962, the Bayliss, Wiley cycle accessories factory at Tyseley was closed, and the building it occupied was sold. Two years later, after Renold had assisted the Board of Trade to find a new tenant, the Abercrave factory was vacated, thus completing the Group's exit, apart from chain, from the cycle fittings industry.

Perry Engineering Ltd. having become fully integrated with Renold Chains Ltd. in 1964, the former was placed in voluntary liquidation in March 1965. It was then possible to reorganize production at what continues to be named Perry Works, Birmingham, as part of a general move involving Coventry Chain Works and described in a later chapter. In the spring of 1967 extensive rearrangement of the Birmingham factory had been completed.

CHAPTER 3

THE HISTORY OF JOHN HOLROYD AND CO. LTD.

The merger of Renold and John Holroyd and Co. Ltd. was the chief event in Group expansion during the years described in this history, the greatest in scale since the Renold-Coventry-Brampton merger of 1930, and the one that went furthest in fulfilling the Renold Board's move into wider fields of positive drive transmission, differing from but complementary to those occupied by precision chain. The outcome of the association between Renold and Holroyd in one financial, managerial and technical whole caused the Renold companies to become a large technical product Group of a specialized kind, diverse but with a common motivation and enhanced strength and influence in the engineering industry. The merger was a triumph also for business diplomacy and technical co-operation between the two parties although the negotiators were fortunate from the outset in possessing advantages accruing from co-operation in the days of their independence. These were reinforced by personally friendly feelings among the principals of both companies.

In 1964, the year of the merger, John Holroyd and Co. Ltd. looked back upon more than a century of successful and growing enterprise, having begun as a public company, become a private one, and finally a public company in 1936. The man who gave the business its name, John Holroyd, was, as far as we know, one of those Victorian engineering designers and craftsmen of that high order for which Manchester became famous, whose archetype may perhaps be found in Sir Joseph Whitworth (one of whose nephews was once a Holroyd director), and among whom, in the course of time, Hans Renold came to eminence.

Little is known about Holroyd personally, and it is many years since anyone bearing the Holroyd name has been at the head of affairs in the company. But John Holroyd appears in a trade directory of 1871 as a sewing machine maker in Hulme, Manchester, and sewing and knitting machines were the goods he mainly produced. That he was a man of resource and inventiveness is suggested by his designing and making a machine for bowling cricket balls. A picture of this apparatus survives in an old catalogue. John Holroyd describes it as being 'so constructed as to combine the various styles of bowling practised by professional cricketers, namely, slow, medium, fast, with a break from the in or offside as desired. With the help of a boy to operate it gentlemen possessing a suitable room or plot of land might obtain excellent practice and become expert batters without the expense of a professional bowler.' Nothing is known of the number made nor by whom they were bought, but it has been confirmed that one machine was used at the Lancashire County Cricket Club ground at Old Trafford, Manchester.

John Holroyd's business grew, and towards the end of the 1880s he was seeking more capital, possibly because some London customers had been unable to meet their obligations to him after he had completed their order for 100,000 children's sewing machines which they subsequently failed to sell. For an engineer in a comparatively small way this must have been a severe setback. Holroyd's difficulty became known to the then owner of the Broughton Copper Works, a Mr Sumner. The latter was approached for advice by a Mr Liebert. Mr Liebert was the principal of a firm of export merchants he had established in 1836, which continued its independent existence into the 'thirties of the present century. At the time of Holroyd's financial problems, the Lieberts lived at Victoria Park, Manchester; there were the parents, four daughters and three sons in their twenties. These three young men came ultimately to control John Holroyd and Co. Ltd. and laid the foundations upon which its later success was built.

Of the three Liebert boys, Harry was the eldest. Like many other men who have done well in life, he was of a practical turn of mind and was neither brilliant nor diligent as a schoolboy, although well endowed with brains and the spirit of adventure. He had no desire to follow his father into the merchant's world, but wanted to be a mechanical engineer, and was accordingly apprenticed to Beyer Peacock and Co. Ltd., the locomotive builders, at Gorton, Manchester. Harry Liebert completed his indentures with credit, had opportunities to ride on the foot-plate, and until recent times no boy would have asked more of life's opening years. From Gorton he was appointed as maintenance engineer of a dredger fleet in the Baltic. His father, however, thought the post had no prospects; he also wished his son to be nearer home. As a result, he consulted Mr Sumner, the only engineer he knew well, on the possibility of a local engineering appointment for Harry.

An introduction to Holroyd followed, and it was in 1887, during one of Harry's home leaves, that the Holroyd financial difficulties were solved by the formation of a company named then, as now, John Holroyd and Co. Ltd. The capital was £9,000. Mr Emil Liebert senior was the Chairman, and Harry became one of three Managing Directors, the other two being John Holroyd himself (whose salary was £300 a year) and Sir Joseph Whitworth's nephew, who was paid, as was Harry, £200 a year but who left, to be succeeded by George Liebert, the Chairman's second son and his partner in the family export business. The Holroyd company began to diversify; machine tools and small tools made their appearance as the main product, knitting machines taking a minor place. Sewing machine manufacture was given up. It is a historical coincidence of interest that the rising cycle manufacturing industry of the Midlands provided an important outlet for Holroyd machine tools.

A catalogue of the year 1888 indicates the scope of Holroyd machine tool development and calls attention to the fact that: 'the tools made by the company have been specially designed to combine the best features of English and American labour-saving appliances'. It covers a wide range of machine tools, a number of textile machines, a variety of cutters, tools and gauges, and even a number of domestic appliances. Among the machine tools are universal, plain and hand milling machines.

Several machines for making screws are listed and separate machines for slotting and polishing the heads; there are lathes built in several sizes and single- and multi-spindle drills. Power presses were a speciality; there were a 'grinding lathe' with covered slides and a tapering attachment, and sawing and slotting machines. Some of the machines appear to forecast future lines of development. There is, for instance, an 'automatic compressed tablet making machine for medicated lozenges', which is described as entirely self-acting and workable either by hand or power. One of the present-day activities of Holroyd is the manufacture on a large scale of the Manesty range of presses and tablet-making machines (q.v., below).

The screw cutting machines were later developed into a thread-milling machine by Harry Liebert, the original of which is in the Science Museum, South Kensington; its success led eventually to the company's interest in wormgear drives.

Among the large number of cutters and tools, special mention is made of 'cutters for the teeth of gear wheels which can be sharpened by grinding without changing their form'. There are also 'epicycloidal patent cutters' which are claimed to cut 'perfectly interchangeable gears'. This may have been the beginning of the company's subsequently large gear manufacturing activities.

Twist drills were manufactured; the catalogue claims that 'these are less expensive than the common drill, and entirely dispense with the constant repair at the smithy fire'.

For the electrical industry there is a 'dynamo electric machine with improved brush holders which run entirely free from sparking'. Then come a number of textile machines, mainly for knitting, sewing and pressing. It is plain that aids to machine-shop engineering were becoming, or had already become, the company's main products.

The time had now come when George, the new Chairman, and Harry, were to be joined by the youngest of the three Liebert brothers, named, like his father, Emil. The last-named, who came down from Cambridge a Senior Wrangler, was an electrical engineer in the service of Mather and Platt Ltd., another great Manchester firm of engineers. He was working on the electrification of the Stockholm suburban railway system when in the summer of 1894 his father asked him to join George and Harry. He agreed, bringing with him a useful reinforcement of engineering experience and intellectual attainment, the more welcome because about this time the original John Holroyd had left the scene, although whether he had died in harness or had retired is unknown.

The business outgrew its premises in Manchester, and by 1896 was established on the site at Milnrow, Rochdale, which it still occupies, although on a much larger scale. The heart of the Milnrow works was the former Perseverance Mill, bought from a textile company in liquidation; the mill was in fact two mills, adjoining which, and part of the property, was a single-storey building, the former schoolhouse of the first school ever to have existed in Milnrow village. It was a source of pride to Mr Liebert senior, who negotiated the purchase, that he was able to acquire the whole of the buildings, including two steam engines, two Lancashire boilers, line shafting, pulleys and belts, together with (as he reported to his colleagues with satisfaction) 'all the gas fittings', for £1,610. The floor space thus made available to Holroyd was six times larger than that at Great Jackson Street.

Two years after the three brothers had begun to take affairs firmly into their own control their father died at the age of 84. As they assumed their new responsibilities they looked out upon an industrial world in which mass production was making its appearance, and with it the need for specialized single-purpose machine tools. Of these John Holroyd and Co. Ltd. had already built a number for the Singer Sewing Machine Co., and had supplied screw millers to the Birmingham Small Arms Co. for the mass production of rifle components. While supervising the production of a number of fine-thread milling machines for Singer and BSA, Harry Liebert realized that no manufacturer had produced a machine for milling coarser threads on longer screws, of the kind used for jacks, tilting wagons, or feed mechanisms (e.g. the lead screws of a lathe), and for manual use to open large windows, fanlights, and so forth. In each case the thread had to be cut on a lathe. This involved 'setting-up' the lathe, which called for the work of a skilled toolsetter, and subsequent turning by a skilled turner, whereas Harry Liebert wanted a single-purpose machine for the automatic production of variously-shaped threads, of which more than one at a time could be operated by unskilled or semi-skilled labour. About 1897, therefore, he designed, built and patented his first long thread-milling machine with a three-edged cutting tool. It was a success, and marked a clear advance on previous practice; the prototype in the Science Museum was followed by many others. By 1906, machines had been bought by sixty-five different firms all over the world, and thirteen were in use in the Milnrow works itself. Its historical importance is the greater because it led the way to the development of machine tools for cutting worms. Hence it stands in the direct line of ancestry to the wormgear drives, forming the principal Holroyd product at the time of the merger with Renold and since. The wormgear drives for speed reduction were the basis of the earlier association of the two companies, Renold finding that the Holroyd drives and their own precision chain drives were complementary in the market, and in a number of countries implementing this technological alliance by selling Holroyd wormgear units through the Renold organization.

It was not long before Harry Liebert's invention led to the introduction at Milnrow of machines specifically designed for worm milling. These were supplemented in 1904 by others for cutting wormwheel teeth; from 1906 gear cutting appears as a distinct production category. Then came machines for grinding worm threads after hardening.

In 1902 the company was reorganized, becoming for all practical purposes the property of the three brothers. It had by then developed a high reputation for precision engineering, establishing it firmly among its customers at home and overseas. It grew with the growth of the engineering industry as a whole, drawing its strength from the increasing demand for machines of greater productivity and accuracy. Turnover rose; more space was needed. This time Providence came to the aid of Perseverance; a former woollen mill, the Providence, adjacent to the Milnrow works, was acquired in 1907.

1907 was a year notable also for a step towards integrated

production, which took Holroyd into the foundry business. The Holroyd company needed a large tonnage of castings. They therefore bought an ironfoundry belonging to John Petrie and Co. Ltd., Whitehall Street and Rope Street, Rochdale. The Petrie company was established in 1814; Renold Ltd. possess the wages book of 1814 among other documents of venerable antiquity. The foundry's main product was steam driving engines for the cotton mills. At the beginning of the present century demand was declining, probably because of competition from the gas engine, and the machine shop depended upon orders for repairs and maintenance. The Holroyd directors thought of reviving engine building. For this purpose they called in a Swiss expert in steam-engine design. Only two horizontal engines, well abreast of the best contemporary practice, were built. But the day of the reciprocating steam engine for industrial driving was nearing its end; in consequence the engine department reverted to repairs and maintenance only, until it finally closed in 1925.

One of the Petrie engines has returned to its birthplace, and is honoured with a building to itself, where frequent visits are paid to it by connoisseurs of what was best in Victorian engineering craftsmanship. This old prime mover is a single-cylinder beam engine with an original nominal rating of 20 h.p., built by Petrie in 1841 for John Hurst, Whitelees Mill, Littleborough. It cost him £650, including the boiler. The Whitelees Mill was later owned by the Co-operative Wholesale Society Ltd., and the engine ran, virtually unaltered, for over 100 years until 1942. It came to Holroyd as a gift from the CWS. It is to be shown working at low speeds, with power supplied by an electric motor taken through a standard Holroyd wormgear speed reducer and a Renold chain drive to the beam motion.

Although the foundry continued to produce iron castings for machining in the Holroyd shops, it included also a small bronze and brass foundry for casting bearings, glands and fittings for steam engines, and other jobbing work. The basic techniques furnished precisely the knowledge and experience needed for casting the bronze for wheels. From this comparatively insignificant sub-unit, then occupying what is now the directors' luncheon room, arose a large non-ferrous foundry, later to make a massive contribution to phosphor bronze founding with centrifugally-cast components, using a process under the trade name 'Holfos'.

In consequence of the war of 1914–18 new buildings and extensions arose, including another machine shop and a canteen. Harry Liebert, fertile as ever in invention, became active in the management of a national shell factory at Rochdale, where he designed and built a system of cranes and conveyors light enough in operation to permit the handling of heavy shell bodies by women munition makers.

By this time growing business had led to the installation of overhead cranage in one of the shops for building heavier machines, including machine tools for locomotive and carriage builders, printing press manufacturers, and shipbuilders. The extending range was accompanied by continuous growth in the gear department, which foreshadowed the ultimate predominance of wormgears as the chief product. The main source of wormgear orders in the 1920s and early 1930s was the automobile manufacturing industry, the principal makers also ordering many machine tools. One of the best customers was the Ford Motor Company Ltd. after the new Thames-side works at Dagenham, Essex, was opened. A visit to Dagenham was almost a monthly occurrence for Harry Liebert, who designed many special machine tools for Ford, including continuous milling and multi-station types.

Three distinct lines of advance are now discernible: (*a*) the machine tools; (*b*) worm and other gears; and (*c*) casting non-ferrous materials, mainly bronze. Much was due to Harry Liebert, but an equal share derived from George's business ability and the less conspicuous but able Emil, not the least of their tasks being to direct some of Harry's more unusual ideas into practical channels.

In 1936, an eventful year for Holroyds, the company had nine Directors, including four younger ones: Mr E. O. Liebert, son of Mr George Liebert, who had joined Holroyds in the late 1920s; Mr F. T. Stott, an able engineer who joined the company as a boy in 1914 and who returned after the First World War, during which he had a distinguished career in the Royal Flying Corps; Mr Clement Meek, a metallurgist who came in 1921, and Mr G. W. Taylor, a chartered accountant, who joined in 1932.

In the late 1920s and early 1930s, Holroyd, in common with others in Britain and the USA, as well as in Europe generally, had a difficult time, and at one stage came almost to the point of extinction. Reference must be made here to Mr E. George Liebert, a son of Mr Harry Liebert, who became General Manager in the 1920s, and who had spent some time at Leyland Motors, where he studied production methods. He left Holroyd in 1934 to become, subsequently, a Director of the Henry Simon Group, but has always been a regular visitor to the Milnrow Works and has maintained his interest in the family business. While at Holroyd, he tackled with great energy a good deal of reorganization and introduced systems and methods which up to then had been lacking. His contribution was an important one during those difficult years.

The five Directors in 1936, in addition to the four already mentioned, were the three brothers Liebert—all, at that time, aged 70 or more, Mr V. Gartside, a machine tool designer and engineer, and an authority on wormgears, and Mr A. G. Bossier, nephew of the Liebert brothers and a linguist, who had charge of machine tool sales. Also on the staff was Mr Harry Walker, an engineer in charge of gear design, who was to become, as Dr Walker, well known and highly regarded in the gear world as one of the foremost authorities on wormgears.

In June 1936, the decision was taken to convert the business into a public company and place on the market about 60 per cent of the company's shares by way of private placing, which would enable them to be quoted on the London and Manchester Stock Exchanges. At the same time the five last-mentioned Directors retired from the Board, leaving the four first-mentioned, i.e. Mr E. O. Liebert, Chairman and Managing Director, Mr C. Meek (Rochdale), Mr G. W. Taylor (Milnrow), Joint General Managers, and Mr F. T. Stott, Works Director and Chief Engineer. This Board remained unchanged for twenty-eight years, although the status of its members altered, as all four became Managing Directors. During this period the business grew and flourished on the foundations laid by the earlier generations.

Some of the technical aspects of wormgear design, manufacture and use and of the wormgear's association with precision chain will be discussed in the next chapter; it should be noted here, however, that the wheel and worm in the speed-reducing gearbox present an efficient method of transmission between a fast-running prime mover and a slow-moving part; the gearbox occupies little space and in one pair of gears can provide the speed reduction required up to, say, 100:1; its applications cover mechanical driving of nearly every kind. In this field Holroyd became pre-eminent. In 1931 they achieved a world record of 98·2 per cent efficiency with one of their gearboxes under test at the National Physical Laboratory, Teddington, Middlesex.

Holroyd's contribution to the national effort during the Second World War had its interesting individual touches. Every Rolls Royce Merlin engine produced during that time was fitted with magneto driving gears made from spuncast Holfos J.H.17 bronze. Over 2,000 machine tools, large numbers of other precision gears and thousands of tons of castings were supplied. The gears included those required not only for the Merlin but, among others, for the production and development of radar and for predictors and locators. In 1943, as a supplement to the necessities of war production, they acquired an old-established company of forgemasters, John Mountford and Co. Ltd., Manchester.[1] By the end of 1964, the year of the merger, their subsidiaries included also Billetop and Gilpin Ltd., Syston, Leicestershire, makers of bearings and bushes from Holfos bronze castings; Sibleys (1922) Ltd., London, also bearing makers, and Manesty Machines Ltd., Speke, Liverpool, makers of tablet-compressing machines and ancillary equipment. In addition, licences to make wormgears had been granted to three companies in America, India and Japan.[2]

[1] Now responsible to the Chain Division of Renold Ltd.

[2] The American company of De Laval-Holroyd Inc., Trenton, N.J., was formed in 1957 in collaboration with De Laval Turbine Inc. The Holroyd interest later passed to De Laval solely, although De Laval continue as licensees. The Indian licence is with the New Allenberry Works Ltd., Calcutta, and the Japanese licensees are the Kobe Steel Works Ltd.

The acquisition of Manesty Machines Ltd. illustrates the unexpected turns a company's history can take. About seventy years ago, Harry Liebert had built to customers' requirements machines for compressing pharmaceutical tablets. The manufacturing chemists, Thompson and Capper Ltd., Liverpool (now part of the Eucryl Group Ltd.), conducted a small engineering section to serve their business, but much of their plant and equipment was obtained from other suppliers, including Holroyd. In 1946 the Thompson and Capper engineering section (known as Manesty Machines simply because it was founded in premises in Manesty Lane, Liverpool) was sold to Holroyd, who continued it as a subsidiary. The commercial significance of tablet-making machinery is not easy to overestimate in the present era, when medicines prescribed in tablet form have increased immeasurably in number, because of both medical discoveries and extended health services, and when sweets are eaten by all and sundry on a scale previously unparalleled. The clinical aspects of this fascinating theme are not for these pages, but the engineering requirements of the tablet-making machine in its different forms were well within the scope of John Holroyd and Co. Ltd. The earlier models bear no comparison with the present types: 'formerly the cams, gears and other moving parts were exposed, but the mechanism is now enclosed, wormgears have replaced spur gearing, plain bearings in many instances have given way to ball and roller bearings, lubrication is controlled and stainless steel has replaced many parts previously made in tinned iron, brass or copper.'[1] That is to say, the machines display what are in many respects features of machine tool evolution.

General expansion made more building necessary at Milnrow. This was completed and in operation in 1962, when new construction had begun at Rochdale also. Shortly before, in 1959, the first and at that time inconclusive discussions on a Renold-Holroyd merger took place. Although nothing came of the talks then, the views of the two companies were converging.

[1] Arthur Little and K. A. Mitchell, *Tablet Making*, Liverpool, The Northern Publishing Co. Ltd., 2nd Edition, 1963. Those wishing to go deeper into the subject will find it fully explained in this book.

Holroyd were now a successful and prosperous group, with a business having assets valued around the £5 million mark, and employing over 2,000 workpeople. In 1962 they had a group profit of over £1 million before tax, the result of an increased output made possible by the new buildings and equipment. There was a moderate rise of 2½ per cent in the total dividend distribution for 1961–2, together with a bonus share distribution made possible as a result of capitalizing £500,000 from reserves—a step taken simply to maintain the issued capital on a scale matching the growth of the organization.

It was natural that Renold should consider a closer relationship with Holroyd; the two companies held each other in high regard. Besides the respect engendered by their equality in engineering skill and inventiveness, their products shared common ground. In the market they tended to make the running one for the other. Their union in a single organization would clinch this association to their joint advantage.

The Holroyd Board had also to think about another pressing question. The Chairman, Mr E. O. Liebert, son of George and grandson of the first Chairman, was considering retirement from office as one of the four Joint Managing Directors. So too was another of these four, Mr Clement Meek, and the retirement of a third, Mr Frank T. Stott, was approaching. It was true that there remained an able senior colleague, the fourth of their number, Mr G. W. Taylor, and that younger men of solid technical achievement in senior positions were available. But without exception all those upon whom would depend the guidance of Holroyd in a further independent existence were deeply and increasingly absorbed in specialized tasks of detailed management, sufficient to call for the entire individual energies of each. The interim solution adopted was for Mr Liebert to remain as Chairman, and an enlargement of the Board. The new directors, however, also had exacting departmental responsibilities—Dr H. Walker as Technical Director (Gears), Mr John Meek as Director for the Holfos works, and Mr H. G. Hiscocks, who had just succeeded Mr Taylor, as Managing Director of Manesty Machines, with Mr Charles Meek, an industrial metallurgist, brother of John (both are sons of Mr Clement Meek, already mentioned) awaiting promotion to the Holroyd Board. In all these circumstances, unification with their old friends, in a complementary industry twenty miles to the south, who had from their earliest days paid close and scientific attention to this very problem of managing a large and complex group organization, was not without attraction. Furthermore, as the Holroyd leaders already knew from the tentative discussions in 1959, the Renold men were ready enough to co-operate. Even so, although more purposive talks began in 1963, no satisfactory basis for a merger was found until July 1964; in that month, however, both parties decided that the time had come.

By the terms of the merger, Renold Chains Ltd. offered to acquire all the issued share capital of John Holroyd and Co. Ltd. The offer was unanimously recommended to the Holroyd shareholders by the Holroyd Board.

The parties had respectively the advice and assistance of J. Henry Schroder Wagg and Co. Ltd., who made the offer on behalf of Renold, and Samuel Montagu Ltd., who commended it to the Holroyd shareholders.

The letter to the Holroyd shareholders set out the benefits of the merger in the following terms:

> 'Renold is already associated with Holroyd in the selling of gear products in many countries of the world. The proposed merger will not only develop this association but will represent a further expansion by Renold into other transmission products with positive drive characteristics and continue the policy of the Renold Board initiated a year or so ago with the introduction of sprag clutches to provide additional overall strength to the organization in the transmission field. These additional transmission products will be made available through the strong and extensive Renold national and international selling organization as well as through Holroyd and Holroyd agents overseas. This extension of product range within the transmission field will enable still further improvements to be made in the servicing of both Renold and Holroyd customers for their transmission requirements. In other words, the existing range of Renold products and the addi-

tional Holroyd range resulting from the merger are complementary and will produce additional sales each for the other.

'It is anticipated that opportunities will result for co-operation in the research and development field which, in due course, will confer benefits in the more widely constituted activities of the combined Group.

'It will be a source of added strength to the combined Group that Holroyd activities are not confined to gear products but extend into the fields of machine tools, including machines for the manufacture of their wormgears; phosphor bronze castings and finished machined components; forgings; and machinery, including tablet compressing machines for the pharmaceutical, chemical, confectionery and other industries.

'It is expected that the result of this merger will be to increase efficiency, and to spread costs for the benefit of the combined Group.'

No difficulty was encountered, and the transaction went through smoothly. Mr Liebert remained as Chairman of Holroyd, with Mr Taylor as sole Managing Director, and there was an interchange of directorships between the two companies. The identity of Holroyd was preserved. It was a union imparting managerial, technical and commercial strength to both parties. On October 27, 1964, an extraordinary general meeting at Renold House passed the resolutions necessary to bring the arrangements into effect by the creation of sufficient new shares and their conversion into stock.

Speaking at the annual general meeting of John Holroyd and Co. Ltd. on January 14, 1965, Mr Liebert said: 'And so we join with Renold, after a peak year and with a full order book, in what we hope and believe will be a merger which will bring benefit to both companies. The identity and character of Holroyd, which is so well known in engineering circles, will be preserved, based as it is on long traditions and experience.' He remarked also on the possible uniqueness of the fact that the same directors—himself, Mr Taylor, Mr Clement

Meek and Mr Stott—had managed the company's affairs for virtually the whole of its life since 1936, when its shares had first been quoted as those of a public company on the London and Manchester Stock Exchanges.

In the following March, Sir Charles Renold defined the intention behind the merger in his annual statement to the Renold shareholders. It was, he wrote, 'the marriage of two precision engineering products of the highest quality in the positive drive transmission field so that each product could complement the other. The further intention was that the Renold Group's world-wide selling organization should provide additional outlets to match the ever-increasing demand for gear products resulting in greater service to customers and increased efficiency in operations. Furthermore, outside the transmission field the other interests of the Holroyd Group provided worthwhile contributions to profits and spread of activities. I am very happy to tell you that although much remains to be done before we can reap the full benefit of this merger, we are settling down together very happily with great confidence in our joint future. We expect that next year and even more so in the years to come, we shall see the benefits of this important step in our development.'

A problem confronting the united companies was a heavy backlog of uncompleted and undelivered orders for Holroyd gears. This initially restrained the Group sales organization from the deployment of its now greatly increased powers to sell Holroyd products. Even so, additional capacity at Milnrow and Rochdale was beginning to make a contribution to output. By November 1965 a new two-storey 38,000 sq. ft. extension to the East Works at Milnrow, designed to centralize the storage of parts and assembly of wormgear speed-reduction units, was in use. A further phase of the expansion programme at Milnrow soon began, bringing about the centralization of all the plant used in machining rotors and making more room available for gear products. The rotor machinery was designed and built by Holroyd and is housed in a new building with a floor area of 19,000 sq. ft., which came into production in August 1966.

By the end of the first full year in the Renold Group, stockholders had reason to be satisfied. The Holroyd contribution to the operating profit was nearly £1½ million. The other Holroyd subsidiaries all showed progress. In Sir Charles Renold's annual statement for the year ending April 3, 1966, he drew attention also to the new acquisition's export contribution:

> 'Holroyd have, for many years now,' he stated, 'directly exported a large percentage of their output of machine tools and rotors, of tablet compressing machines and ancillary equipment, as well as wormgear speed reducers and loose worms and wheels. During the financial year no less than 45 per cent of the sales of these three product ranges were directly exported, and a large share of home sales was also exported indirectly as component parts of finished machines exported by Holroyd customers. The other product ranges of Holroyd which are mainly non-ferrous castings and forgings are not readily exportable but, even including these, Holroyd exports were more than 28 per cent of total sales during this financial year. These exports were shipped to almost every country.'

The speaker's reference to rotors touches upon a fascinating region of Holroyd activity. In 1967 the Queen's Award to Industry was conferred jointly on Renold and Holroyd—upon the former for export performance and the latter for technological innovation. The technological innovation was in respect of milling machines and related equipment for the production of screw-type helical rotors. The rotors produced on the Holroyd machines are the main components of the modern compressor, which would not have been an economic possibility had it not been for the manufacturing knowledge and experience developed by Mr Stott.

Basically, screw-type compressors comprise male and female intermeshing helically-lobed rotors housed in a chamber formed by the rotor casing. The rotor shafts are supported on bearings and are usually phased by timing gears. As the rotors turn, air is drawn axially into the space between them and the casing. Further rotation cuts off the inlet port, trapping the air and compressing it by the screwing action as it is forced towards the outlet port and discharged.

Since the rotating elements of the screw compressor may be kept from contact with each other by means of external timing gears, internal lubrication is unnecessary, so that gases can be compressed without contamination. It is, of course, extremely important that the clearances should be as small as possible if reasonable efficiency is to be obtained. This demands great accuracy, and was not achieved until many years had been spent by manufacturers and engineers in attempts to make rotors economically.

In 1950, when Mr Stott and Dr Walker (later, as we have seen, to become Technical Director, Gears, at Holroyd) were visiting Stockholm, they were introduced to the patentees of the SRM (Svenska Rotor Maskiner A/B) rotor, from whom they learnt of the efforts to make it chiefly by contour planing, which always had to be followed by highly-skilled hand work. As a result of the discussions, Holroyd concluded that milling would be better than contour planing. Trials were made on existing Holroyd machines, with results so encouraging that Holroyd undertook to build a precision machine which would eliminate the need for hand work. A machine to cut rotors of up to 20-in. diameter was made for the British licensee of the SRM compressor, with a guarantee that rotors of up to 12-in. diameter would be produced directly from the machine and inside the required tolerances. The cutters required for larger diameters were so much bigger than had been used before that their behaviour could not be forecast. It has since proved possible to finish rotors without handwork up to 32 in. in diameter, largely by employing both a method for producing the profile form of the cutters on the machine itself and a machine for grinding the cutter profiles.

Five years after the original talks in Stockholm, Holroyd were able to offer all the equipment necessary for manufacturing the rotors, thus enabling the SRM compressor to be made economically by licensees in many countries. The next step was to the larger compressor, and in 1962 Holroyd delivered the first of a larger series of milling machines to Japan. In 1964 the first fully automatic machine was exhibited in London. There are now upwards of twenty SRM licensees in both hemispheres, all of whom are equipped exclusively with Holroyd machines. The technique has subsequently been applied to cutting hydraulic pump screws and flow meter rotors.

The Queen's Award, incidentally, was received in the year of Mr Stott's and Dr Walker's retirement, crowning the fine engineering achievement of the two men whose work had done so much to earn it.

The newcomer to the Renold Group was the largest it had ever absorbed in its history. At the time of the merger, Holroyd was a group within the Group; that being so, integration was necessarily beset with more problems, but by the end of 1967 the objectives of the merger were being achieved.

THE LIEBERT BROTHERS: (left to right) GEORGE, HARRY and (below) EMIL

CHAPTER 4

ASPECTS OF POSITIVE POWER TRANSMISSION

Positive power transmission means in simple terms the transmission of power without slip, and consequently with maximum efficiency. Positive drive by chain and gears will ensure that power in the form of horsepower put into a transmission will emerge to drive the mechanism concerned with only fractional losses in between. Chain drive efficiency is 98·5 per cent and that of gear drives only a little less. It is contended by the 'positive', i.e. chain and gear school, that drives relying on friction tend to slip and that transmission power losses can be considerably greater than with 'positive' techniques. Therefore, they claim, not only does positive transmission operate more efficiently and accurately, but since power costs money, it also operates more economically. Sometimes savings in power costs alone will pay for a positive drive in as short a time as twelve months when compared with other forms of driving. The customer needs quality, efficiency, accuracy and economy. This is what the Group aims at giving him, and the Renold conviction is that this can best be done with positive power transmission as the criterion.[1]

The Renold–Holroyd merger made it possible to market more effectively a positive drive having a versatility and an applicational range matching those of precision chain. It also made easier the close combination of gearbox and chain which so often provides a technically sound solution of driving

[1] There are, of course, circumstances in which drive slip is desirable, particularly where there is a risk of accidental overload; nowadays this requirement can be met by retaining the attributes of positive drives but fitting friction discs or torque limiters.

problems (Fig. 1). These results are illustrated in these six typical examples:

Drive for a waste disposal unit, by 5·0-in. 70:1 ratio Holroyd gearbox, through a Renold disc coupling and a 6:1 ratio Renold stock drive;

The rolls of an aluminium rolling mill, driven by 1-in. pitch triplex chain from two Holroyd gearboxes coupled by a Renold chain coupling;

Sack elevator, employing two strands of Renold 8-in. conveying chain of 24,000 lb. breaking load, driven through a Holroyd gearbox and a Renold stock chain drive, a Renold flexible disc coupling connecting gearbox and motor;

Holroyd double-reduction gearbox with a double-extension output shaft coupled into the line shaft of an automatic welding plant by Renold flexible disc couplings;

Motorized brick transporter (with winch): in this, the electric motor drives the (underdriven) Holroyd speed reducer by 0·50 in. pitch simple (single-strand) chain; a 1-in. pitch duplex chain from the output shaft powers the winch and wheels;

The push dog conveyor in a kiln, driven by a train comprising two Holroyd wormgear reduction units coupled by a Renold flexible disc coupling and three Renold chain drives employing 0·375-in., 0·75-in. and 1·0-in. pitch chains.

The technical affinity between chain drive and wormgear is summed up by Renold in a private Group paper of 1965, by Mr J. N. Perryer, the Overseas Sales Manager, who remarked that:

'Although power transmission may involve a speed increase, or no speed change at all, the vast majority of applications demand a speed reduction. Most prime movers function at high speed, a necessity if horsepower per unit of volume is to be reasonable and if prime mover cost is to be economic. There are relatively few driven mechanisms which operate at equivalent or higher speeds—exceptions lie in the field of such things as centrifugal pumps, fans, generators and grinding tools. Most final motions, however, are relatively slow in comparison to the speed of the prime mover, and hence the need for speed reduction in the power transmission medium. A single-stage chain drive is limited, for practical purposes, to a ratio of 6:1 and Holroyd standardized worm boxes (Fig. 2) to 70:1. One chain stage and one wormgear stage will therefore provide up to 420:1 (or an output of about 3·5 r.p.m. from a normal 1,420 r.p.m. electric motor). A double-reduction worm unit can provide a ratio up to 5,000:1, so with only one or two drive stages a range of speed reductions from 1 to 5,000:1, which will cater for most applications, is available. An extension to three or even four stages is possible with very little increase in space utilization. With a first-stage chain drive, a worm unit and a final chain drive, ratios up to 2,250:1 can be obtained and by using a double-reduction worm unit with a chain stage on both sides this can be increased to the enormous figure of 180,000:1.'

In the following pages, brief accounts are given, first, of Renold precision chain development and application with particular reference to the years covered by this book, and then of the inception and development of the Holroyd wormgear.

PRECISION CHAIN IN MODERN ENGINEERING

Although the Renold Group has now made its way into power transmission engineering beyond the manufacture of precision chain and chain accessories, continued development of chain has not been neglected. Faith in its merits burns as brightly as at any time in the past. Progress continues, because of both the intrinsic merits of chain drives and the stimulus of competition from other power transmission systems, for example, belt drives, gears and hydraulics.[1]

The four basic claims for the chain drive are its 'positive' (i.e. non-slip) quality, space-saving, elasticity and reliability. In a privately circulated Renold publication these advantages are described as follows:

[1] The term 'hydraulics' as used in reference to engineering purposes is nowadays applied to fluids other than water, e.g. oil; alternative expressions are 'hydrodynamics' and 'hydrostatics', both in common engineering use.

'Chain drives will function at centres almost as short as for gears and as long as for belts while chain sprockets are smaller in diameter and less in width than belt pulleys transmitting the same horsepower. Chain offers a greater choice of centre distance than either v-belts, toothed belts or gears, thus allowing greater flexibility in positioning. In addition, the adaptability of chain to gear on both sides allows considerable latitude in the disposition of the driving and driven units. From one driving shaft more than one other shaft can be driven and reverse drives can be accommodated by means of a suitably disposed jockey wheel.

'The lubricating oil film in the chain joints, combined with the catenary effect of the free length of the chain, gives a beneficial measure of elasticity. When driving such large units as calendering rolls there will be no "chatter" marks on the paper or cloth as would be the case with gear drives.

'Roller chains provide a form of transmission unaffected by atmospheric or climatic conditions. The desired speed ratio between the driving and driven shafts is always maintained. Slip cannot take place and chain drives can deal with heavy starting loads. Finally, the reasonably well-engineered chain drive will give many years of dependable service.'

These features were summed up also by the Chief Sales Engineer of Renold, Mr L. Jones, in his paper 'Modern Chain Drives', presented to a conference on design engineering and management organized in 1966 by the Production Engineering Research Association of Great Britain, in these words:

'A system of mechanical power transmission that is positive, yet flexible; that can be adapted to a wide range of centre distances; that is commercially available as a standardized product; that has as high a mechanical efficiency as any other form of mechanical power transmission.' Later in the same paper he examined the generous limits governing the use of chain drives. 'In British practice,' he stated, 'the 8-mm. chain will transmit 3 h.p. at 6,000 r.p.m. while at the other extreme four strands of triplex 3·0-in. pitch chain will transmit 4,000 h.p. at 300 r.p.m. This means that from the smallest office machinery to the largest dredger pump and large generators, a chain drive is the effective answer for transmitting power.'

RESEARCH

During the past ten years, research and development have been directed to the improvement of existing products and processes of manufacture and the introduction of new ones, either peculiar to the Renold company or adopted from generally known and accepted advances in metallurgical and engineering practice and not exclusively of company origin. This follows the pattern of advance established with the introduction of the bush roller chain in the 1880s. During subsequent years discoveries in metallurgy have led to the evolution of commercial steels and other metals having improved chemical compositions, and to the use of new alloying elements. In the same period industrial chemistry and the skill of plant builders have made possible a number of improvements in heat treatment processes, while the engineer has contributed ideas and inventions that have revolutionized machine shop practice. These have been progressively introduced in the works of the Renold Group, there to be supplemented by the results of development inside the organization, making full use of Renold-owned facilities with the aid of both fundamental and applied science.

In 1958 a new Research and Development block came into use at Renold Works, Manchester. The R & D work was done through two departments, one concerned with the investigation of processes used in production, the other with the development of different varieties of chain and chain drive accessories. Staffed by a handful of able men to whom the Renold reputation owes much, it was nevertheless ceasing to keep pace with events. Between 1955 and 1962 the R & D activity was therefore extended to provide sections covering product design, experiment, metallurgy, and internal and external liaison, with a new process investigation section. The resources of the R & D department were increased again in 1960 by the erection of an internal combustion engine test house for investigating the behaviour of camshaft drives on automobile engines, lubrication and associated engineering problems. In the test house, one of the most up-to-date in the world, there are nine test beds, and service is provided for the automobile industry generally, in addition to the testing and development of Renold products. The work done has helped to make the Renold name internationally eminent in automobile camshaft drive design, and famous car manufacurers go there for help. In addition, there are, of course, a chain test house and a metallurgical laboratory.

The extent and intricacy of the company's and its customers' engineering needs have made increasing calls on the department, which has travelled far since the twofold system of the late 1950s and before. Responsibilities have grown with the various phases of expansion in the United Kingdom and abroad. At the present time, the growth and importance of the department have led to the provision of still wider facilities to be housed in new buildings, now under construction, on a site next door to Renold House.

CHAIN PRODUCTION PROCESS DEVELOPMENT

The persistent rise in the cost of labour and materials has stimulated the movement towards greater mechanization in processes as the expense of handling components between operations has gone up. To meet this situation, a forward process-mechanization policy has been introduced, extending an already sophisticated machine shop practice. The new ideas rely largely on metal *forming* as distinct from metal *cutting*, and are given effect by a considerable extension of the use of presses as alternatives to, or reinforcements for, metal-cutting machine tools.

The bush roller chain consists essentially of inner and outer plates, bushes, rollers and pins. Manufacturing methods based on the principle of forming rather than cutting have been or are being applied to all these components of each chain link, some on a pilot and others on a production scale.

In the same connection, air cushion support (i.e. on the 'Hovercraft' principle) has been employed to eject components to die level, and many components are cold-extruded. Some of the dies, incidentally, are being made with success in a machinable tungsten carbide which can be annealed, altered and hardened again without distortion.

Automatic assembly is used in the final stages of manufacture, testifying to the accuracy with which the many thousands of separate parts are made, since the advantages of the techniques can be obtained only with high quality and dimensional correctness in long runs of components.

From the point of view of process development, in a perfect world material should come in at one end of the works, and finished chain leave at the other without the manual intervention of an operator. This will come to pass, although at present it is impossible to say when. On that day, full automation will have been achieved.

Renold Works, Manchester, and the Coventry Chain Works have each shared in the company's contribution to the use of powder metallurgy; the Coventry works was at one time, and possibly may still be, the largest user of Swedish iron powder in Europe. The process technique is the compacting of finely-ground powdered metal, which is then heat-treated by sintering, thus binding the minute grains in a single mass.

During the past twenty-five years powder metallurgy has gained much ground. Its potentialities had long been known to metallurgists and engineers, but it was during the Second World War and after that it came to be applied on an industrial scale. Despite the inevitable problems, it is now established for the production of many small engineering components. It also has the advantage that non-metallic hardening or lubricating substances can be mixed with the powder. This feature provides a form of 'self-lubrication' because of the absorption and retention of the lubricant by the moulded powder in its finished shape. Given the solution of the metallurgical problems, sintered powder has obvious attractions in the production of long runs of some chain components, and is therefore being increasingly used by Renold for chain rollers of certain sizes.

The fact that tungsten carbide, well known to engineers as a source of extremely hard cutting tools, is capable of being powdered and compacted, has led to the use of powder metallurgy in the Renold Works, Manchester, toolroom for making dies for use in some of the presses.

The toolroom itself, where these dies are only one among many other products used in works service, has been much enlarged in recent years. It has always been a remarkable engineering establishment and stands, as it were, at a junction between scientific research and process and product development. Its origins and traditions go back to Hans Renold. In his toolroom he set up a standard of craftsmanship of the most exacting kind, which is still rigorously upheld. In 1955 the toolroom gave employment to 300 craftsmen in an area of one of the main works blocks. In its latest form it is the workplace of 400 modern successors of the craftsmen who were the pride of the engineering industry in its earlier days. Now housed in what is called the 'new service building', the erection of which began in 1961 and was completed in 1962, it includes, in addition to toolmaking for the company and its constituents, the works maintenance services, and a bay where the patterns are made for the wheel and machine-frame castings bought from foundries for machining, finishing and assembly.

Even so brief a reference to production processes as this would be incomplete without mention of the gradual introduction into the Renold Chain Division's works establishments of numerically controlled machine tools, the development of which has been studied by the company for many years, although they first made their appearance in the Renold shops in 1964–5. At the present time, their usefulness arises from the fact that it is becoming harder to find skilled and versatile operators for standard machines. These, when used infrequently for batch work, require setting-up time and craft skill. The time, however, becomes increasingly costly, and the skill grows scarcer. The company is therefore seeking a solution in the tape or similarly controlled machine. An example would be the manufacture of a special wheel, where a punched tape or plug board pattern would enable the machine-operating instructions to be filed in readiness for intermittent requirements, the

Fig. 1. Combined chain drive and gearbox

Fig. 2. A Holroyd gearbox (sectioned)

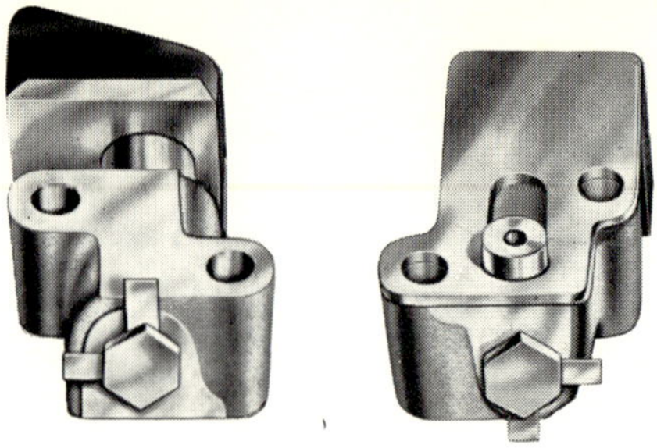

The patented Renold SCD adjuster

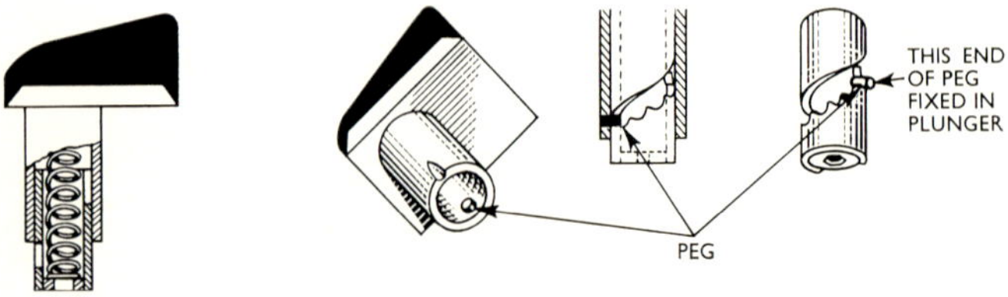

The non-return restraint mechanism

A typical camshaft drive
fitted with SCD adjuster

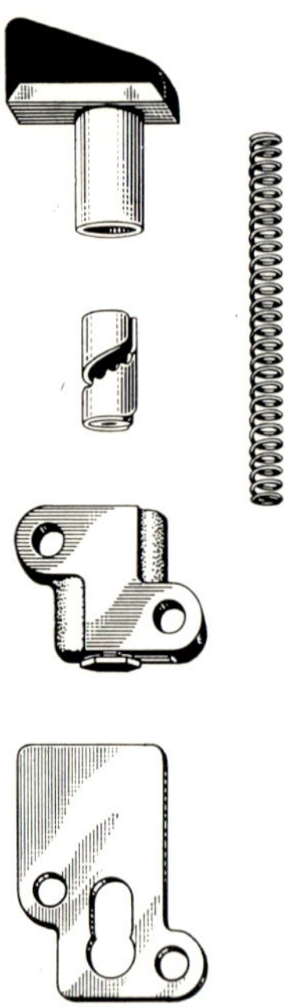

The parts of an
SCD adjuster

Fig. 3. The SCD adjuster described in the text

Fig. 4. The Sprag clutch

Fig. 5. Employment of chain in an automobile main drive

tape and tool arrangement details then being all that would be needed when the demand arose.

Renold do not look to this kind of control for large-volume production; in their view, the present generation of fully automatic machines serves equally well for this purpose and meets it more efficiently.

CHAIN PRODUCT DEVELOPMENT

Historically, the most notable product developments to become established commercially by Renold Chains during the earlier part of the period under review have been the SCD adjuster and the sprag clutch. The latter is particularly noteworthy, for it marks one of the major steps in giving effect to a deliberate departure from precedent in the policy of product diversification which, while complementary to chain driving, would go beyond chain driving alone.

The SCD (Short Centre Distance) adjuster was evolved for use in a specific application, namely automobile two-point camshaft drives. It was first announced in 1954, and was the outcome of development work which extended over the two preceding years.

In the normal course of its service a chain will wear, like anything else. Adjustment then becomes necessary in the majority of drives. This is achieved on industrial drives in one of three ways—by a direct increase in centre distance, by making one wheel shaft moveable, e.g. on an electric motor, or by incorporating a jockey pinion on the normally unloaded strand. With some applications, notably internal combustion engines, however, automatic adjustment is preferable. In 1927 the company introduced the Renold adjuster, which consisted of a jockey pinion mounted on a slideable bearing, which together were forced into the chain by a spring, thus taking up any chain slack caused by wear. Return motion was prevented by pawls engaging in serrations. This adjuster was used on the 0·375-in. pitch chains, while for 0·5-in. pitch the Coventry adjuster was patented in 1932. The latter again used a jockey pinion, but in this case on an eccentrically-mounted bearing, the bearing being turned by means of a clock-type spring. Return motion was prevented by a pawl and ratchet mechanism similar to that incorporated in a bicycle freewheel.

The Renold and Coventry adjusters were somewhat bulky. With the reduction in size of automobile engines it became necessary to design a smaller and more compact kind. In 1954 Renold introduced the SCD (Fig. 3), comprising a rubber slipper head bonded to a steel plate and mounted on a plunger. The plunger slides in the bore of the adjuster body and inside the plunger is the restraint cylinder. The slipper head is forced into the chain by a combination of spring and oil pressure and return motion is prevented by the limit peg (fixed to the plunger) engaging with one of the serrations in the spiral slot in the restraint cylinder. This adjuster is widely used on automobile engines in the United Kingdom and in continental Europe. There are two basic models; adaptations to each make it suitable for specific applications.

By 1956 250,000 SCD adjusters were in use, and the device had already been adopted as a standard fitting on a number of British motor-vehicle engines, including those of the British Motor Corporation 1·2, 1·5 and 2·6 litre types, the Jaguar 2·4 and 3·5 and the Standard 2·2 diesel. Its introduction strengthened the position of the 0·375-in. pitch chain as the almost exclusive method of driving camshafts on United Kingdom-manufactured passenger car engines; its use, now reckoned in millions, is general and it has penetrated deeply into overseas markets, including the Japanese.

It is indeed remarkable, and an encouraging reminder that there is another side to the alleged lack of enterprise shown by British manufacturers in international markets, that Renold and their agents in Japan in 1966 secured the exclusive supply of camshaft chain and SCD adjusters for the Toyo Kogyo 'Luce' car, for which a successful future, both Japanese and international, seems assured.

In other technical categories, the company introduced in 1955 a range of stock pinions, finished-bored and keywayed to standard shaft sizes. This was rapidly extended, both as to size and bore, with provision for metric bore sizes when desired.

Another development of the same period was the introduction of standard stock wheels with one-sided bosses, for conveyor chain, enabling the chain pull to be taken as close to the bearings as possible by fitting the boss outwards. Alternatively, by fitting it inwards, unusual layouts can be accommodated, for example, when the chain centre line is required to be at the ends of overhung shafts; again, when used in conjunction with swing-tray elevators, head-wheels mounted on stub shafts with the bosses inwards to the bearings provide greater clearance for the trays.

1956 saw the preparation of designs for heavy conveyor chain giving service up to a breaking load of 300,000 lb., and the redesign of the AS (American Standard) chain series. Chains to American Standard dimensions were first made by Renold during the 1939–45 war for use in American equipment; since then the company has steadily introduced marks to cover the greater part of the American range, certain pitches of AS chain now being made up to octuplex widths.

In 1957 the company introduced a range of sprag clutches made under licence from the Formsprag Company, Michigan, USA. The Formsprag clutch resulted from a military requirement during the Second World War for an overrunning clutch of great capacity and small dimensions. 'Overrunning' in this context meant a clutch permitting an aircraft-engine supercharger drive to continue functioning under its appropriate power loading when another drive was put out of action. Kindred and numerous applications in industrial use and for other mechanical purposes were soon envisaged. Originally the clutches were imported, but now nearly all of them are made in the United Kingdom, at Renold Works, Cardiff. By 1959 it was possible to establish twelve categories of service in which the sprag clutch (Fig. 4) could be used with advantage for seventy-two different applications where overrunning, backstopping and indexing are required.

The basic principle of the clutch is the sprag, which may be described as an unusual design of cam operating between two races. In one direction, one race will overrun or rotate relative to the other, and in the opposite direction it will drive. Energizing springs are used to hold the sprag in engagement with the two races, thus eliminating backlash. When driving begins, the engaged sprags drive instantly. A retainer holds the sprag independently and permits it to take up its proportionate share of the load. It also holds the sprag in such a position that during the overrunning phase of the cycle a minimum force is exerted on the inner race cam, hence reducing wear. The design can be compared to a pendulum, of which the lower end is the lower section of the sprag. The sprag's ability to drive is dependent upon its gripping angle, the kind of force exerted being similar to that of a stick pushed against the ground. If the stick is allowed to make an acute angle with the ground, it will slip; if the angle is much nearer the perpendicular it will not, but will attempt to deliver torque to the surface on which it is acting. Two years after the agreement, the Renold version of the clutch had become a stock line enabling off-the-shelf service to be given. Improvements in design now yield higher rates of indexing and higher speeds in overrunning. Moreover, with growing experience, Renold are able to offer a great diversity of standard types.

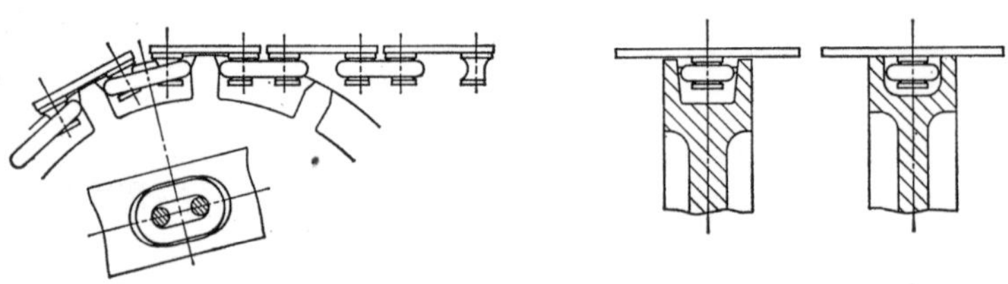

Fig. 6. The multi-planar link chain: operation in the vertical plane

Three years later, in 1962, an entirely new product appeared, the multi-planar link chain for conveying (Figs. 6 and 7). The chain consists of stainless slats which can negotiate corners in the horizontal plane, an attribute of considerable advantage when conveying bottles, especially flasks which do not easily accommodate themselves to the turntables frequently employed with slat band chain. At the pitch points of the chain, two grooved studs are welded to the underside. Each slat is connected to the next by a single zinc-plated mild steel cable chain link fitting in the groove of one stud on each slat, the groove allowing sufficient vertical clearance to permit articulation in this plane also. A flat on each stud enables the cable link to be disconnected by swivelling the slat, making for ease of installation and adjustment without tools. Driving is by toothed pinions at selected points in the chain run, with idler pulleys at points other than those used for driving. The chain is a contribution to the food, bottling and pharmaceutical industries because of its simple design, which makes for hygiene, and its strength, permitting its use for all normal requirements of circuit layout corner drive units.

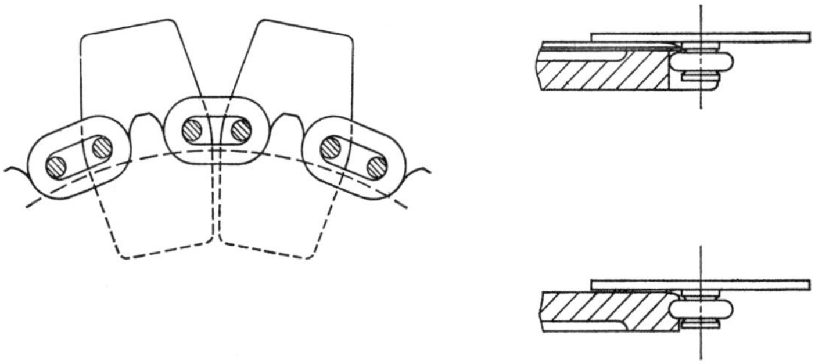

Fig. 7. The multi-planar link chain: operation in the horizontal plane

Brief reference may be made to two other product developments nearing the full-scale production stage. These are (1) self-lubricating chain, and (2) torque limiters. The former incorporates a self-lubricated sintered (powder metallurgy) bush. It is for use where working conditions and positions prevent lubrication and where limited life is therefore acceptable. The torque limiters, for use in conjunction with plate wheels,[1] are designed to slip at a predetermined torque, thus

[1] 'Plate wheels' are normally of uniform thickness throughout (generally corresponding to tooth thickness); they can be of either split or solid design, and are suitable for bolting or otherwise securing to separate hubs or flanges (themselves secured to a rotating shaft). Split plate wheels offer easy assembly or removal in applications such as shunting locomotive axles, where removal of a conventional (solid) type wheel would also involve removal of rail wheels.

protecting both the chain and the machine by preventing accidental overload.

PRECISION CHAIN APPLICATION

Precision chain application is of enormous variety. Since chain first became an effective means of driving, it has been taken up by every industry, the word 'industry' here being given the widest sense it will bear. Each year seems to offer further openings in fields already developed, and to them is added the prospect of other areas where new driving needs seem to find their optimum solution in the use of chain, a simple enough product in itself, but presenting fascinating, intricate and difficult problems in engineering.

The market opportunities are vast. The Group's catalogues already list more than 1,000 individual chain and ancillary products, and the additions to the range consistently outnumber those items abandoned either because of obsolete and obsolescent uses or, in the course of range rationalization, because of sheer unprofitability.

What the customer wants may not, and ordinarily does not, call for completely new driving concepts. Most needs are met by efficient adaptation from stock chains, wheels, pinions, clutches and attachments as an outcome of consultation between the user and the Renold engineers; there have been and continue to be scores of interesting ways in which Renold products have been successfully incorporated into the equipment and machinery of most industries. These, it is true, are applicational developments, but rather as a variation on the theme of their use to suit a particular requirement.

Although the company is busy satisfying demand, it is also creating it. To be ready with the applicational answer when change occurs in designs, for example, in camshaft drives (both automobile and marine), nuclear plants, and safety devices in conveyors, and to have the technical answer to the operational difficulties that arise, is fundamental, and creates new growth points.

With these general considerations in mind, let us look at a

few of the applications between 1956 and 1967; they are not exhaustive but they are typical, effectively illustrating the kind and the extent of the contribution that chain driving makes to industry.

Mechanical handling and the use of conveyor chain present an example. The search for greater production efficiency involves more mechanization of both material and product handling. The applicational development work in this region covers the formulation of new chain product requirements, e.g. overhead conveying and product modifications to meet the conditions in particular industries, such as the multiplanar link chain mentioned in the previous section.

In the conveying and mechanical handling field, the scope for chain and chain attachment applications is virtually limitless. The company's steel roller conveyor chains are marketed in twelve standardized breaking strengths from 3,000 lb. to 85,000 lb. in a variety of pitches. They incorporate hardened steel bearing pins and bushes, and rollers at each pitch point to minimize friction when carrying a load and when engaging the teeth of the wheels. The two main types are manufactured with either solid bearing pins for strength and weight-carrying, or hollow bearing pins for lighter duty and to permit the insertion of spigot pins, stay-bars and other carrying and handling attachments. The range of standardized attachments is large—a selection of thirty of them is illustrated in a company publication of 1964.

The bush roller chain itself can be used for conveying purposes proper as well as power transmission; an example is the inserted block chain patented by Renold in 1928, in which blocks are inserted between the inner plates and the required attachments are fitted to the blocks. This use of bush roller chain is illustrated in the Lapointe horizontal continuous broaching machine for machining the bolt side faces, joint faces and half-round bore at car-engine connecting rod big ends. The basis of the machine is a pair of special 3·5-in. pitch bush roller chains with hollow bearing pins running continuously between fixtures to hold the connecting rods between them and carrying the rods past the broaches. The hollow pins enable the fixtures to be removed or fitted without dismantling the chains. The maximum length of the broaches and thus of the cutting stroke is 120 in. and the chains, each having a breaking load of 100,000 lb., may achieve a combined pull of 20 tons.

Recently, a range of conveyor chains of greater width, known as the Mark 2, has been introduced in metric dimensions.

The Royal Navy has found a use for chain in diesel-electric paddle tugs. The paddle steamer may have vanished from the high seas and from the pleasure-trip runs between seaside resorts, but the Navy still finds that paddle propulsion gives useful service when tugs are needed to manoeuvre large warships like aircraft carriers in confined spaces. The tugs are 157 ft. long, with a loaded displacement of 710 tons. The four diesel-electric generating sets supply power to two 800-h.p. motors, each driving an independent paddle shaft by chain. Although the shafts can be coupled if the tug is required to go to sea, for normal harbour service they are independently driven. The speed reduction is 8:1, the paddle shafts turning at 35 r.p.m. to attain normal cruising speed. These are the largest Renold power transmissions ever supplied for use in seagoing vessels, and although the application is not new, it is interesting as the Navy's choice in the most modern, if not the last, functional role of the paddler.

Paddle propulsion at sea links us with the end of the sailing ship and the earliest days of steam. But marine oil engines followed steam, and considerable applicational developments have occurred in this area during the 1950s and 1960s. This has been partly due to the increasing size of engines, and partly to the employment of chain drives from propeller shafts to generators and auxiliary pumps. The capital cost of such drives is high, but the savings in operational expense outweigh it, because several hundred horsepower can be taken from the propeller shaft with little effect on the main engine consumption, compared with the needs both in fuel and maintenance of independent auxiliary engines. Incidentally, when the ship docks, the passengers who are landed and the cargo that is discharged may well be dependent on diesel locomotives for

An automobile engine test cell at Renold Works, Manchester

An artist's impression of the new Research and Development Centre, Wythenshawe; Renold House is on the right

further transport, and these too will embody camshaft and auxiliary driving chain.

Supposing, however, that our passengers or consignors of cargo decide to use the air instead. Chain plays its part in many of the world's aircraft; a case in point is the Comet 4, now proved and established, which employs chain for elevator and aileron trimming and in the flap system.

While on the subject of aircraft, a special 'one-off' application in aircraft manufacture, adopted by Short Brothers and Harland Ltd. at Belfast, deserves mention because it shows how chain's versatility makes it useful in odd and unexpected ways from which the Renold Group will never make a fortune but which fill a need and illustrate the resourcefulness of the chain engineer. The requirement was to provide for uniform and speedy wrapping of the glass-fibre ducts used to carry the heating, ventilating and electrical services in aircraft. The machine to do the wrapping consists basically of a rotating circular ring upon which are reels of glasscloth tape. Through the centre of the reels passes a former on which the tape is wound at uniform speed. On the outer periphery is a gear ring of 0·5-in. pitch roller chain secured by attachments at every ten pitches. The pinion engaging the chain is driven through a four-speed gearbox providing variable ring speeds. The former passing through the centre ring is supported from a profile board, by the side of which is a rack consisting of 0·5-in. pitch chain driven by a pinion through a reduction gearbox, and serving the purpose of indexing the amount of overlap of the tape wrapping.

There is a substantial use of chain in machine-tool design. Consider a multi-spindle vertical automatic, the 'Verticalauto', for turning, boring, facing and similar machine-shop operations. This machine has eight chucks, and one of its features is the incorporation of eight externally gearing chain rings, through which the drive is transmitted to the chucks when performing light drilling work. For all other operations, power is transmitted to the chucks from the main drive shafts through hydraulically controlled dog clutches. When any chuck is required for light drilling, its clutch is disengaged and a chain pinion is geared into the chain ring; the chain pinion is carried on a swinging arm forming a unit with a ¼ h.p. hydraulic motor and reduction gear. There are three such motors, so that any three chucks can be driven simultaneously in this manner.

Turning to the roads, in all highway construction chain drives for different purposes are to be found in earth-moving machinery, excavators, crushers, pumps, track drives, mixers, gravel-washers, road rollers and finishers—indeed they run through the whole gamut of contractors' light and heavy plant.

On the roads made with this kind of equipment pass to and fro the millions of the world's motor vehicles. In the majority of British and European car engines precision chain has been used for many years. Today practically all British and many leading European manufacturers employ Renold Group chains; more than 20,000,000 engines have been fitted with Renold timing chains. This is the outcome of the objective engineering advantages possessed by the chain. Because it permits compactness and flexibility in drive layout and centre distance, the designer has considerable freedom in positioning the camshaft to comply with limitations dictated by machining, clearances and accessibility. Centre distance tolerances are relatively large, and the effects of cylinder block expansion on the chain drive are negligible. Long centres are spanned without difficulty, e.g. in the overhead camshaft engine, and the flexibility of the drive matches the needs of the vee or twin o.h.cs. Ample lubrication from the engine oil supply, together with minimal wear ensured by use of the SCD adjuster, obviate troublesome or uneconomic manual attention. The chain may normally be expected to last the life of the engine. The drive is usually located at the front of the cylinder block, and built into the engine with an oil-tight casing, thus needing minimum maintenance and providing safety. In a well-engineered drive the noise level is low and, indeed, inaudible against the ambient noise of the engine itself. At the time of writing the trend in engine design is towards the o.h.c., which has made its appearance in, among many other cars, the Hillman Imp, the Rover 2000, the Opel Rekord, the Mercedes Benz and the Peugeot 204.

Progress in science and technology sometimes leads to the reappearance in an altered form of methods once thought to have run their course. Today chain drive transmissions in motor vehicles (Fig. 5) are enjoying a renaissance which is almost certain to affect automobile engineering for a long time to come. A significant example in its possible future implications is the four-wheel drive of the Ferguson R/52 car, with its new type of centre differential, receiving power from the engine through a step-down chain drive; Renold collaborated in the work which produced this chain application. Further encouragement for advocates of chain drive has been provided also by the independent selection of chain, a selection resulting from the makers' own design studies, as the transmission medium in the front-wheel drive of the Oldsmobile Toronado, although Renold studies of the subject are claimed by the Group's engineers to have preceded the work of General Motors. Chain was adopted in the Toronado because it affords less restriction in centre distance compared with gears. In accordance with American practice, the kind used is of the 'silent' or inverted-tooth type, to which designers in the United States have tended to adhere since Hans Renold's 'silent chain' patent of 1895.

The present-day Renold work on this subject was summarized in *Automobile Engineer* (May 1967, page 180) as follows:

'During the early stages in the evolution of the motor car at the end of the nineteenth and beginning of the twentieth century—precision power-transmission chains were also making an impact, owing to their simplicity, reliability and high efficiency. Naturally, therefore, chain drives to the rear wheels of cars were to be seen in great variety for many years until about 1905, when the cardan shaft transmission arrangement began to supersede the chain drives. But the designers did for the final drive gears what they had never done for the chains: they enclosed and lubricated them!

'With the evolutionary progress of design, the cardan shaft layout is again challenged, as more and more examples are seen of the unitizing of the engine, gearbox and drive to the wheels at one end of the car. The gains in respect of passenger space, styling, traction and handling are well known, so this trend is certain to continue.

'Such arrangements generally demand the transmission of engine power at high speeds from the axis of the crankshaft to a parallel axis in the gearbox. While the first design approach was the embodiment of a gear train, there is with high speed gearing the fundamental problem of noise. Many factors of design and method and precision of production affect the level of audibility. Control of all of these factors is very costly and, normally, a uniformly low level of noise cannot be guaranteed with a high-speed gear train.

'Realizing that this problem existed and that a chain drive offered a ready means of improvement, Renold Chains Ltd. [now Renold Ltd.] began development several years ago—certainly, they claim, before the General Motors work on the Toronado. Experience has also been gained in collaboration with Harry Ferguson Research Ltd., whose requirements for the chain drive for their four-wheel drive car were similar.

'As a result of the programme undertaken by Renold Chains Ltd., the following obvious requirements were met:

compactness—no more space is required than for alternative drives, such as gears;
uniformly low level of audibility;
reliability;
durability;
economic viability.

'To meet the first requirement, a precision roller chain was developed which, for its size, had higher breaking strength and higher fatigue resistance than ever attained previously, and these characteristics were coupled with quiet running. Not only had the chain to be of reasonably small pitch, but also it had to be capable of control throughout its life, that is, automatic adjustment was required. For this particular application the needs are rather special, in that whichever strand is carrying the load—drive on one strand or overrun load on the other—it must run directly from the driving to the driven chain-wheel without being deflected over the adjusting slippers or jockeys, otherwise excessive wear, audibility and heat losses are entailed. Consequently, a double headed adjuster was developed and patented: this not only takes up any slack in the chain but also has the important characteristic of affording hydraulic damping of slipper head movement, so that the change from drive to overrun is entirely smooth and snatch-free.

'Reliability in the sense of quality control of materials and manufacture to exacting specifications is an everyday routine to Renold Chains Ltd. and therefore was no problem, but in the sense of ensuring that nothing would fail or give trouble even under severe abuse from the driver, further development work had to be carried out. So, as part of the programme to ensure that safety factors under extreme conditions were satisfactory, the transmission was equipped with strain gauges and intensively tested with converted cars at the Motor Industry Research Association proving ground. Durability of course can be positively established only by perseverance with many hundreds of hours of running on the test bed and many thousands of miles of road work under every condition of operation. Now, BMC 1100 and 1800 test units and cars are producing results which, happily, show a transmission life commensurate with that of the power unit or gearbox itself.

'As regards cost, the chain wheels are of conventional materials cut to normal tolerances, which are relatively wide compared with those necessary for gears, as also is the tolerance on centre-to-centre distance. It has been estimated that, by virtue of planning for mass production of the chain, adjuster and chain wheels, the precision roller chain drive can be made an economically viable project, as compared with alternatives, and with the bonus of the advantages of quietness and smoothness already mentioned. Moreover, there may be an additional bonus, in the form of higher efficiency, reflected in more miles per gallon; but this depends also on many other parts of the power transmission and is currently being investigated. Certainly, converted cars show a new characteristic of free running, which suggests a gain has been made.

'So far, most of the work has been done on BMC front wheel drive cars, that is, with a positive connection between the crankshaft and gearbox. However, for transmission drives from torque converters to automatic gearboxes, the demands are much less exacting than with the positive connection. Renold Chains Ltd. intend to continue their development work on both types of transmission and also for transmissions on cars other than BMC. In due course, even conversion units may become available. If these developments, augmented by those taking place in other countries, result in the success expected and chain-type primary drives return to automobile engineering, then the well-established merits of the roller chain will have been proved in another important field, with benefits in respect of both performance and service.'

Automobile applications are further demonstrated in the three-litre formula Repco-Brabham, with which Mr Jack Brabham won the Manufacturers' and Drivers' World Championship in 1966. Each of the two cylinder heads has a single overhead camshaft, both o.h.cs. being driven by one 0·375-in. pitch simple chain. The same chain drives the fuel injection and ignition equipment by means of a 23-toothed pinion mounted on the top strand. Tensioning is by one 23-tooth jockey pinion. The necessary 2:1 reduction is achieved in the first stage drive from the crankshaft to a layshaft; a 0·375-in. pitch duplex roller chain tensioned by an SCD adjuster is used in this first stage drive. 'Racing' chains, to meet the extreme conditions imposed by motor car and cycle racing, are used on both stages. Vibration damper strips are fitted to three of the four straight runs of the camshaft chain.

Another recent example is the Triumph-engined 'Gyronaut X-1' motor-cycle, Mr Bob Leppan's mount when he broke the solo motor-cycle speed record at Bonneville salt flats on September 1, 1966, at a mean speed of 245·667 m.p.h. In this machine, two tandem-mounted Triumph Trophy 650 c.c. engines employ Renold 0·50-in. pitch racing chain both for the primary drives and for the drives connecting the two engines. Bob Leppan tuned these engines personally to a claimed output of 110 b.h.p. at 7,300 r.p.m. using an alcohol fuel. The final drive passes this power to the rear wheels by 0·625-in. pitch Renold racing chain.

Applications in automobile manufacture are not wholly concerned with auxiliary drives in the engine or power transmission to the road wheels. Although powered window-winding mechanisms are still comparatively rare, many schemes, both hydraulic and electric, designed to provide the motorist with pushbutton opening and closing, have been tried in the past with limited success. A successful unit has, however, been designed by Rolls Royce in collaboration with Renold and is incorporated in Rolls Royce cars, 155 pitches of 8 mm. pitch chain being used to raise and lower the windows. Quiet, smooth acting, compact and reliable, many thousands of these units have been, and still are being, fitted since they were introduced in 1958.

A no less interesting use is exemplified by the 0·375-in. pitch simple roller chain employed in the door-operating gear of some public service vehicles. This illustrates one of the dynamic advantages of chain transmission—its ability to impart torque instantaneously without fluctuation in either force or velocity throughout the cycle, giving smooth movement in opening and closing the doors, a result harder to obtain with a simple lever action because of the angular movement of the cranks.

It might have been expected that atomic energy would have uses for precision chain, and so it has proved. In 1956 the company began to study the use of chain for raising and lowering the carbon control rods governing output in the pile. These chains are of special design for use in unique circumstances. They are made to obviate the adverse effect of high temperatures and the other working conditions; in the same field stainless steel conveyor chain serves the purpose of removing radioactive effluent.

And so we could go on, were there sufficient space between the covers of one book. Those who wish for fuller information must turn to the abundant literature produced in extremely readable and often non-technical form by the Renold publicity organization. Enough has been said, however, to show the way in which the precision chain enters as an essential component into the tremendous business of supplying man's needs and desires, beginning with the primary ones of food, clothing, shelter and transport, and thence moving on to the consequent refinements, even to the point at which he demands complex mechanisms for the construction of rocket motors with which to probe into space, for chain plays its part here too. It is a far cry from Slater's 'bowl' chain of just over a century ago.

PRECISION CHAIN STANDARDIZATION

Renold Ltd. have been interested in the standardization of chains, wheels and pinions from the early days of the bush roller cycle chain, and later as it made its way in other applications for heavier driving. The precision chain industry's single short-lived experiment with its own trade association had standardization as its principal motive. The call for chain needed in the war effort of 1914–18 had involved this problem, and the principal makers sought to co-operate in solving it upon the initiative of Hans Renold Ltd. In this way four makers found a common ground of association—Renold, the Coventry Chain Co. Ltd., Perry and Co. Ltd., and Alfred Appleby Ltd. (the last-named was subsequently absorbed by Renold Chains Ltd.); the other three, as we know, also merged in the course of time. The then independent companies formed the Association of British Driving Chain Manufacturers. The principal work of the ABDCM was done by its technical subcommittee, but the forces within its ranks were ill-balanced, the more so after the Coventry company seceded, and in the 1920s the Association ceased to exist. However, it contributed in some measure to the establishment of a basis for standardized pitches. The work had its influence on the first British chain Standard (BS 228: 1925) which gave gearing dimensions. BS 228 was subsequently amended in 1934, the amended version including more detail and providing rationalized dimensions for certain chain. The latest version, of 1962, gives details of tooth form and carries the dimensional rationalization a stage further.

After the 1939–45 war, the British Standards Institution's technical committee concerned with steel roller chain and sprockets (under the chairmanship of a Renold man, Dr O. Burawoy, then the company's Chief Development Engineer) proposed the creation of a committee of the International Organization for Standardization (ISO) with the task of establishing international standards for these products. The international body accepted the proposal, and the new committee had its first full meeting in London in 1961, Dr Burawoy presiding. Almost complete agreement was reached on a draft international standard for agricultural chain, and since then the committee has continued, through an international working group, to prepare basic proposals for short pitch transmission roller chains. It is at work also on conveyor and other mechanical handling chain.

In addition to activity in the ISO, Renold have taken part in establishing national standards in all countries where the opportunity to do so exists, through co-operation with the national standards organizations. It has thus contributed to the revision and reissue of BS 228: 1925 and the other British work.

INCEPTION AND DEVELOPMENT OF HOLROYD WORMGEARS

Since 1964 the engineering achievement outlined in the foregoing pages has been extended by the merger with John Holroyd and Co. Ltd. Although the concept of the wormgear is old (Leonardo da Vinci, just as he did with driving chain, embodied the principle in some of his designs), it took many years to overcome a prejudice against it, probably caused by the inaccuracy of earlier wormcutting tools. Even today, when complete dimensional accuracy and high working efficiency have been achieved in Holroyd gears, prejudice occasionally tends to colour the views of engineers, some of whom consider the wormgear to be necessarily inefficient. It is significant, however, that once this initial distrust has passed, the reaction is normally enthusiastic. Holroyd early found this to be so in Belgium, Italy and Sweden: in recent years a similar conservative feeling in Germany has begun to disappear.

It was in the decade 1930–40 that Holroyd entered the wormgear business on a large scale. The two main outlets sought were (*a*) totally-enclosed standardized wormgear boxes for industrial use: (*b*) the automobile industry—practically all manufacturers of goods and big passenger vehicles were then employing worm drives for the rear axles transmission.

Expansion was greatly helped when in 1931 Holroyd obtained a world efficiency record on a pair of their wormgears

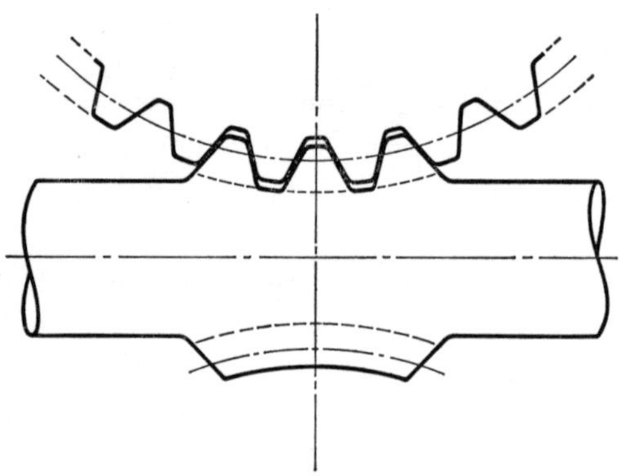

Fig. 8. The Hindley worm

which was tested independently at the National Physical Laboratory, Teddington, Middlesex. It created wide interest and drew attention to Holroyd as leaders in the field.

The history of wormgear efficiency tests began about 1913, when the famous motor car designer and engineer, Dr W. F. Lanchester, was using a worm drive for the rear axle transmission on the Lanchester car, almost the only private car using the drive at that time. It was of the Hindley type (Fig. 8), in which the worm threads partly encircled the worm-wheel, and had been developed by Dr Lanchester. To prove its merits, Lanchester built what was known as the Daimler-Lanchester testing machine, which he subsequently presented to the NPL. In a paper of 1913 presented to the Institution of Automobile Engineers, he described the machine and published results showing efficiencies of about 96 per cent. He wrote a further paper published in 1916. Both caused controversy, arousing criticism among makers of the cylindrical type of drive.

Four years later David Brown and Sons Ltd. made known the result of a test on the Daimler-Lanchester machine in the course of which their own make of cylindrical worm drive gave the remarkable maximum efficiency of 97·3 per cent. This was a distinct advance; the improvement can best be

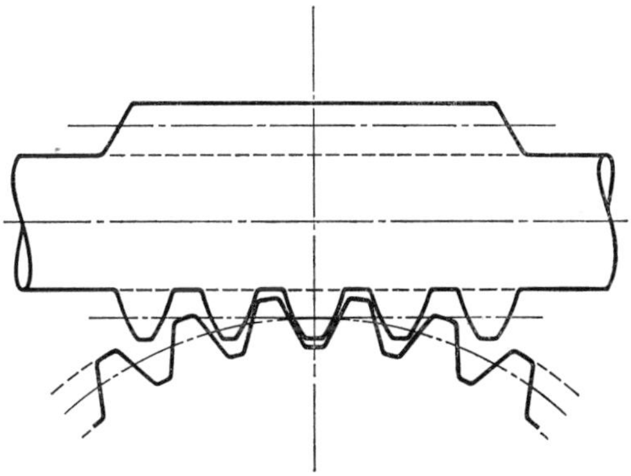

Fig. 9. Cylindrical or parallel worm as made by Holroyd

measured in terms of efficiency loss rather than in actual efficiency, i.e. the David Brown test represented a loss of only 2·7 per cent compared with the previous figure of about 4 per cent.

In 1931 Holroyd submitted a pair of their own wormgears for test at the NPL. The worm (Fig. 9) was cylindrical, but the thread form, the subject of a patent application in the name of the company and Dr H. Walker, differed from that of the conventional cylindrical type. The test gave a result showing efficiencies up to 98·2 per cent, or an efficiency loss of only 1·8 per cent. The grant of a patent was opposed by David Brown, and the wording of the specification had eventually to be confined to a finished product, the process of manufacture being disallowed. In the end, the application was abandoned, Holroyd taking the view that the main point had been obtained by achieving their record. Since the Holroyd tests in 1931, no others have been published, so it is to be supposed that the record is unbeaten.

Dr Walker himself was surprised at the efficiency obtained, and doubted the accuracy of the Daimler-Lanchester machine. After experimental work at Holroyd's, he gave reasons for these doubts in a paper written for the Institution of Mechanical Engineers. Among other things he pointed out that after allowing for the oil churning and bearing losses in the test gearbox, the net wormgear tooth efficiency, calculated from the NPL results, could be up to, and in danger of exceeding, the 100 per cent mark. It could, however, be accepted that although the Daimler-Lanchester machine gave results a little on the high side, it also gave true comparative results between the various gears tested, and the Holroyd record remains unbeaten and unchallenged to this day.[1]

The six chief reasons for the use of wormgears were stated in a private Renold paper circulated in 1966. They are:

(1) All ratios from 5:1 to 70:1 can be accommodated in a single-reduction wormgear. With spur gears the maximum ratio in one pair of gears is limited approximately to 7:1; above this it becomes necessary to double or even triple the reductions, with a multiplication of moving parts. Further, it is possible to accommodate standard speed reductions of 5,000:1 in a double-reduction wormgear, which can be increased on a special basis to 10,000:1.

(2) The wormgear gives a right-angle drive which is often convenient for compact adaptations to a machine.

(3) For a given horsepower, the wormgear is usually smaller in size and more economical than the spur or helical box, particularly above a 7:1 ratio.

(4) Wormgears give inherently quieter and smoother running than do spur and helical gears.

[1] Much of the technical information in this section has been collated from *Reprints from Holroyd newsletters on subjects related to gearing*, 1954; the account of the NPL tests is based upon notes kindly contributed by Dr Walker.

(5) There is less disparity between the comparative efficiencies of different kinds of gears than is generally believed, and in some instances the efficiency of the wormgear may be more favourable.

(6) Designers are increasingly taking advantage of the properties of the modern wormgear, and incorporating it in their work. British engineers use it without hesitation if it suits the applicational design requirements.

The technical aspects of wormgear manufacture are outlined in the Holroyd *Reprints*. They state that in the nineteenth century it was already well known that a multi-thread worm or worm of high lead angle should be more efficient than a single-thread worm. These principles were not, however, widely used because the need for high-efficiency wormgears did not exist and the means for their production were imperfect. Before the advent of the petrol-driven vehicle there was little or no demand for a high-efficiency wormgear and, in fact, the wormgear was used frequently on applications where low efficiency was thought to be an actual advantage because of its tendency to make the gears self-sustaining. Single-thread wormgears with large ratios of reduction were, therefore, thought the best, smaller reduction ratios being achieved by bevel or spur gearing.

But, the *Reprints* explain, the invention of the petrol engine created a demand for high-efficiency drive with a comparatively low reduction ratio for the rear axles of passenger and commercial vehicles: 'Wormgearing was tried and proved highly successful, and this led to the development of methods and machines for providing wormgears of ratios down to about 4:1 or 5:1. The application of wormgears to rear axles is an exacting one, calling for high efficiency, silence of operation and durability under very heavy loads. In developing such gears, improvements had to be made in production techniques, particularly in producing wormgears of higher lead angles and larger numbers of threads than had previously been used. The accuracy and surface finish of the gears were improved and important developments such as casting centrifugally took place in the production of phosphor bronze for the worm wheels.'

Success followed, and it was realized that such small reduction ratio gears applied equally well to other branches of industry. Hence wormgears gradually became used in applications where previously only bevel or spur gears would have been considered. This tendency has continued; more wormgears are now being used in ratios approaching unity, and this has been the main trend. This does not mean that wormgears with the larger reductions are being abandoned: 'on the contrary, the larger reductions are still being used on an increasing number of applications. What it does mean is that the overall useful range of ratios for worm gearing has been extended at the lower end so that instead of confining wormgear ratios between limits of say 10:1 and 70:1, the designer can now choose any ratio between limits of 1:1 and 70:1.'

In special cases there is no difficulty in making wormgears of much greater ratio than 70:1, and these are made to specific orders. At the other extreme, wormgears form a ready means for speed-increasing gears, and in many instances they have been made to increase the speed in ratios up to about 15:1.

With the appearance of this tendency, Holroyd devoted much time, thought and expenditure to the development of manufacturing for close ratio, high lead angle gears, and developed wormwheel hobbing machines, thread grinders, testing machines and hob manufacturing plant for the purpose. There is, of course, a limit to size and power; in general, ratios between 1:1 and 3:1 can be applied to wormgears up to not more than about 8-in. or 10-in. centre distance, and most of the applications fall within about 5-in. centre distance or less. The final decision depends on the speeds and power transmitted.

A typical example of the large-scale application of close ratio is provided by the auxiliary drives for automobile and aircraft engines. Camshaft, pump and distributor drives call for ratios of 1:1 and 2:1 and the wormgear is compact and efficient for this purpose. Numerous examples are to be found in machine tools, particularly on high-speed cutter drives where a right-angle wormgear drive is often most convenient. Among many other industries finding a use for them are the makers of chemical plant, textile machinery, papermaking, wire drawing and high-speed conveyors.

WORMGEAR PRODUCTION DEVELOPMENT

In order to obtain the best performance from a pair of wormgears, the materials should usually consist of a case-hardening steel for the worm, and phosphor bronze for the wormwheel. 'There are three factors which affect the performance of the gears in so far as the combination of materials is concerned,' state the Holroyd *Reprints*. These are:

(1) *Wear:* the ability of the materials to withstand the conditions of the drive without showing appreciable wear after a long period is obviously an important requirement.

(2) *Heat generation:* the amount of heat generated from a pair of wormgears is important because it can affect the running temperature of the oil bath which has to be kept within practical limits. The coefficient of friction between the two materials should therefore be as low as possible in order to achieve a high efficiency, and thus to keep down the oil bath temperature. The steady running temperature of a continuously rated gear is approximately proportional to the coefficient of friction, other conditions being equal. A poor combination of materials may be only a little less efficient than a good combination, but will be found to run at a much higher temperature. For example, a gear which is only 90 per cent efficient would run approximately twice as hot as a gear which is 95 per cent efficient. A low coefficient of friction is therefore important for continuously running wormgears; it may not be important on gears running only for short periods at a time, since the gearbox can cool between runs.

(3) *Strength:* the physical strength of the materials must be such as to avoid the risk of tooth breakage. In general a wormgear usually has a high safety factor in this respect, as

the limits set by wear and heat usually occur long before the strength limit is reached. On gears designed for very intermittent running, however, the limiting strength of the teeth may be approached, because the allowable load for wear and heat can both be stepped up considerably.

So far as the requirements of maximum resistance to wear and minimum heat generation are concerned, Holroyd are satisfied that there is no known wormwheel material giving better results than their Holfos centrifugally-cast phosphor bronze, produced by a process they themselves have developed. It approximates to the bronze defined in BS 421 (*Phosphor Bronze Castings for Gear Blanks*), or in other phosphor bronze specifications closely allied to BS 421.

The method of casting phosphor bronzes to BS 421 is important. The properties to be expected from the three methods of casting (sand cast, chill cast and centrifugally cast) are stated in the specification. The mechanical properties of a centrifugal casting are superior to those of a chill casting, which in turn is superior to a sand casting.

Figures for tensile and compressive strengths do not, however, give a true measure of the wearing properties of a gear material. A material of a high-tensile strength may be inferior in this respect and in coefficient of friction to a material of a much lower tensile strength. For example, a high-tensile manganese bronze wormwheel would wear more rapidly than a centrifugally cast phosphor bronze wheel to BS 421, under the same conditions of loading.

A guide to the wearing properties may be obtained from BS 721 : 1963, *Worm gearing*. That part of the Specification relating to the tooth and thread design of wormgears is largely based on a Holroyd article on wormgear design published eleven years previously in *The Engineer*. Certain design difficulties which arose from the previous Specification were overcome by the Holroyd suggestions.

Comparative allowable surface and bending stress values for the phosphor bronzes defined in BS 421 are given, the actual figures being:

Photo-elasticity test equipment for chain

Glasgow: one of the new-style branch offices and stock depots

	Wear	Strength
Sand cast	1,500	7,200
Chill cast	1,800	9,100
Centrifugally cast	2,200	10,000

The figures are intended to be used in conjunction with the formulae stated in BS 721 for calculating the horsepower capacity of a pair of wormgears. This capacity can be calculated on a basis both of wear and strength, the smaller of the two results being the one to be adopted. The formula in the specification includes other factors dependent on gear speed, length of life of the gears, and combination of materials for the worm and wheel. The figures are therefore to be regarded as comparative factors rather than as actual allowable stresses. It is, however, true that the strength figures are proportional to the ultimate tensile strength of the bronze, and the wear figures approximately proportional to the surface crushing strength of the bronze in lb. per sq. in. or any other units. It is also true that the horsepower capacity of a pair of gears is directly proportional to the figures in the table, e.g. a centrifugally-cast bronze will transmit, for a given life, 46 per cent more load than a sand-cast bronze.

As regards the coefficient of friction, no comparisons are available in this or any other specification, but Holroyd experiments have shown a definite superiority for the centrifugal casting over both the chill and the sand casting.

It may be interesting here to consider the use of high-tensile alloys like manganese bronze and aluminium bronze. On certain intermittently rated or slow-moving wormgears the power that can be transmitted may be limited more by the strength of the material than by its wearing properties. If a gear is used only occasionally and operates at low speed, there is less opportunity for wear to take place and consequently the loading can be increased to the limit set by the necessity of avoiding actual tooth breakage. (Examples of such applications are traction winches, lock gate operating gear, and screwdown mechanism for rolling mills.) The question therefore arises whether in such instances it may be advantageous to use material of a higher tensile strength than centrifugally-cast phosphor bronze for the wheel. Holroyd have concluded that only rarely is it necessary to adopt anything stronger than Holfos centrifugally-cast bronze.

Coming now to the worm or wormshaft, the requirement is for a hard surface of an appropriate core strength. Holroyd use a nickel case-hardening steel. A case-hardening mild steel is avoided because of the risk of surface cracks after grinding the thread surface.

It is emphasized that there are many applications in which the size of the wormgears is dictated by practical considerations. In a machine tool, for example, the centre distance of a pair of wormgears may be determined by other considerations than those of load-carrying capacity; it may be desirable to adopt a larger size of gear than would otherwise be necessary in order to fit in with other parts of the machine. If so, cost comes into it, since load capacity is of secondary importance, and it seems wasteful to use expensive materials if cheaper ones will do. In such circumstances the worm may be left unhardened, and made from a steel of about 0·5 carbon and chill-cast or sand-cast bronze may be used for the wheel. Cast iron is sometimes employed for light duty wheels but does not wear so well under sliding conditions. A steel worm mated with a steel wheel would invite scuffing and seizure, even at slow speeds.

In conclusion it may be said of wormgears that if there are no restrictions on design, the least expensive results are obtained by designing the gear to a minimum size from the best materials. If materials of lower wearing qualities are used, the sizes must be increased, and the overall gain is problematical.

The foregoing consideration led Holroyd many years ago to adopt 'chill casting' whereby the outer surfaces of symmetrical castings are cooled rapidly. With a worm wheel casting, the depth of chilling does not usually exceed the depth of the teeth, although, of course, some beneficial effect is felt right through the casting. For a plain surface component, such as a bearing, the chilling may be less.

The registered name of 'Holfos' bronze, as first known, was the outcome of a later development of the above method, and provided a speciality casting for wormwheels, and later for bearings in the form of bars, tubes, small gear blanks, liners and similar components.

Probably the most far-reaching technical development in Holfos history was the introduction of centrifugal casting in the 1930s. The centrifugal process as such is not peculiar to Holroyd; it is used extensively, for example, in the production of horizontally spun cast-iron service pipes for gas and water mains. As a result of their original research and their contribution to this technique, it became possible for Holroyd to apply a centrifugal casting method to the production of what became known as Super Holfos Bronze, in applications for gear blanks, bushes and other circular castings. At Holfos Works it is employed vertically for casting hollow phosphor bronze bars. The action of the spinning mould, inducing fluid pressure, combined with the chilling as used when casting in static moulds, brought about a grain refinement in the metal that added to the strength and density of the product. The company developed their patented method of combined centrifugal and chill casting for hollow blanks and tubes under the trade mark Spuncast. The process was so successful that it was licensed for tube drawing to ICI, Thos. Bolton, and Vickers. Other modern foundry techniques are used at Holfos besides centrifugal casting, and in addition to the supply of wormwheel and other components to Milnrow, much contract business is done outside the Group, as the foundry is able also to finish-machine the castings, a service appreciated by the customer.

Holroyd have always maintained a large machine-tool design and erection department; this has helped them to provide themselves with machines, tools, and testing equipment specially designed to meet their own individual requirements in the production of wormgears. It may be said that every operation on the tooth and thread production of Holroyd wormgears is carried out on plant of their own design. The works is at present self-contained also in testing and research facilities, and has its own foundry laboratory.

* * *

The relationship of Renold chains and Holroyd gears, notwithstanding the previous selling arrangements, was incomplete until the merger. Now, however, it is possible for a single supplier to put forward a positive power transmission package combining mechanical efficiency with flexibility of application, long life, and freedom from breakdown, with smaller space requirements than a number of other transmissions, and to make it available in most parts of the world, in many cases from stock. This position has, of course, been enhanced by the entry of the Croft companies into the Group.

An early effect of the new phase in positive power transmission opened by the merger was the completion in 1966 of negotiations between the English Industrial Estates Corporation and Renold Chains Ltd., as a result of which the corporation is constructing a large works for Renold on the Tees-side industrial estate at Thornaby, Yorkshire. The initial development consists of 150,000 sq. ft. of factory buildings on an eight-acre site, where from 600 to 800 people will be employed. Renold have, however, reserved enough additional land greatly to extend the unit when necessary. Production of many chain and gear products is expected to begin at Thornaby at the end of 1968.

CHAPTER 5

EXPANSION, RATIONALIZATION AND MARKETING, 1956–67

In 1957, demand was outstripping the productive capacity of the company's various works. At the outset of the period covered by this book the prospects were good despite expected but, as it proved, temporary, recessions. In a number of cases, trade in manufactured goods had leapt upwards: for example, in round terms the world's motor cars increased in number by more than 40,000,000 between 1948 and 1958; the global production of electricity and steel extended similarly. In order to match resources with opportunities, therefore, the company entered upon vast measures of expansion. These provided, in terms of sales, for marketing on a scale 125 per cent greater than in the financial year before the war.

In all this, a cardinal problem was the rationalization of existing capacity and the provision of new. In 1955, conveyor chain was made at Renold Works, Manchester, but the resources to produce it had to be doubled; something of the same kind was happening at Coventry Chain Works with the rising demand for chain needed in agricultural machinery added to the growing calls upon Coventry in other directions. Hence it was decided to provide a separate works for conveyor and agricultural machinery chain at Burton-upon-Trent, Staffordshire.

Premises at Burton formerly occupied by a firm of enamel manufacturers were acquired in 1955, together with some adjoining land. This gave the company approximately seven acres for a new factory and eight to allow for subsequent growth. Burton-upon-Trent is a tightly-knit and somewhat self-contained community, so the Renold immigrants made themselves known in the district at an inaugural lunch on January 26, 1956, to which were invited the mayor, representative civic officials, and other prominent citizens.

Plans were prepared for a new factory, with offices at the front enveloping the existing premises, and the opportunity was taken to ensure that, as far as possible, the whole factory layout could be arranged for rational movement of parts from raw material at the north end to finished chain and dispatch at the south.

Agreement was reached with the previous owner that his employees should be given the opportunity to work with Renold. Approximately ninety joined, from shop-floor labourers to senior staff. Initially, they were kept occupied by the expedient of bringing from Coventry such work as could be conveniently done by hand on improvised benches. Others were engaged in making alterations to the existing property in preparation for the new building. Sections of the new factory were taken over successively from November 1957. The first of the office blocks came into use in July 1958, but it was not until early in 1959 that the final works section began to operate. The official opening, by the then Parliamentary Secretary to the Board of Trade, Mr (now Sir) John Rodgers, took place on May 7th. During the course of the project, certain key Renold personnel were transferred from Manchester and Coventry in order to form a nucleus of experienced employees. Apart from this, workpeople were locally recruited and were trained either on the premises or by short spells in Renold works elsewhere.

In October 1957, and in co-operation with the trade unions, the Renold system of joint consultation was introduced at Burton. A works 'at home' was held in June 1959, arousing lively interest among employees and their families. For many of them, it was their first insight into the magnitude of the enterprise which employed them and of the ends to which their efforts were directed. Since those pioneering days conveyor chain production at Burton has consistently increased, and is now three times greater than that of the original department in Manchester. The diversity of the product has increased significantly. Whereas the Burton works at first made one basic series of conveyor chains, a variety of special kinds is now produced.

The company made provision also for an increase in automobile chain capacity, and for doubling the resources for heavy roller and cranked link chain output, although these of course affected Renold Works, Manchester, and Coventry Chain Works, not Burton. Altogether, the expansion programme up to this point cost a total in round figures of £2,500,000.

There was a similarly vigorous movement in sales and marketing. In Britain during the early 1960s much thought was given to branch office, stock depot and distribution arrangements for United Kingdom sales. The company decided to re-site and re-style the Leeds branch, and this established the pattern for many other branches. To provide better service, sites were found close to ring roads, preferably near main road junctions, main line railways, and airports. The number of these new-style buildings has grown steadily, and all the branches have been rehoused in the past ten years.

It is convenient here to emphasize a feature of the Renold selling method, applied from branches such as these, and overseas as well. The managers and salesmen at home and abroad, who are naturally selected because they possess the human qualities likely to commend them to actual and potential customers, are all by a strict rule men with an engineering training, and the qualifications accompanying membership of the engineering institutions. They do not work on commission. They are established professional men feeling sure enough of themselves to advise a would-be purchaser against using a chain drive if they judge it to be the wrong solution for his problems, or to persuade him to accept a less expensive variant than he may have wanted at first if it will save him money and do the work better. This approach pays. These men sell, and sell successfully, and while the bonus in goodwill cannot be measured arithmetically, it is certainly reflected somewhere in the figures.

Changes in the Renold organization during this epoch occurred overseas as well. Established at different times in many countries, the company's operations outside the United Kingdom extended considerably between 1956 and 1967. Significant steps were taken to increase sales and begin wheel manufacture in Canada. In Europe, either the expansion of old-established works and branches or the creation of new ones affected Austria, Belgium, Denmark, France, West Germany, Italy, the Netherlands and Spain. These, by the way, are mentioned in alphabetical not chronological order.

Chronologically, the first to be noted took place in February 1957, when another link was forged in the long-standing Renold association with North America. That association had been firmly established over half-a-century before: it had passed through many competitive vicissitudes, and sometimes through phases of limited co-operation between Hans Renold and his company and the United States chain manufacturing firms. Consolidation had come when Hans Renold Ltd. began to export and distribute their British-made chains effectively in the Canadian and American markets after the First World War. But 1957 claims attention as something more than the reinforcement of existing arrangements. What took place was a departure, at once technical and commercial, from the past policy of the exclusive, or nearly exclusive adherence to chain and chain drives, characteristic of the company from the beginning of Hans Renold's career as a master man and a manufacturer in business on his own account. The sign of change, and of the coming of a new epoch, was the agreement with the Formsprag Company, of Warren, Michigan. By it, Renold Chains Ltd. were licensed first to sell the Formsprag industrial clutch (the technical applications of which have already been mentioned) in the United Kingdom, and then by stages to manufacture sprag clutches in the Group's British factories. The significance of the agreement was its clear indication of a readiness to accept product diversification, a policy that began to take effect in 1956, has not yet reached its limits, and of which the future ramifications and their effect on the Group's product range are impossible to forecast with exactitude at the present time. Product diversification was to be extended on a much wider basis by the merger with John Holroyd and Co. Ltd. seven years later; nevertheless, the Formsprag-Renold agreement was a decisive step in giving substance to the company's growing conviction that it might well widen its interests in positive power transmission. Commercially, Renold broadened their range with a proven accessory of a kind complementary to chain driving. Furthermore, the licensers, the Formsprag Company, had designed and built a clutch that was in its own right a worthy companion to that range. It is precision engineered, is supplied primarily to the designers and makers of the plant and machinery in which it is to be incorporated, and is sold on its quality and at its fair price. All this ensured close identity with the Renold outlook upon both business and engineering.

The departure from the 'chains only' principle was inevitable once industry had become convinced of the advantages of chain driving. Even when 'chains only' was the working rule of the company, the chains were soon provided with wheels, the chain clutch was evolved, other chain-driving accessories on the borderline of 'non-chain' products were added, and special tools for servicing and maintenance were designed and supplied to chain drive users.

On June 14, 1961, the works of a new Canadian subsidiary, Renold Chains Manufacturing Ltd., Brantford, Ontario, was formally opened. Since Renold moved into Canada after the First World War, the company's business there has prospered steadily on the sale of British-made chain drives, and the Brantford factory was built and equipped for the purpose of making wheels and pinions for use with chains made to American Standard specifications, exported to Canada, and sold by the selling subsidiary created after a reorganization of the company's Canadian interests in 1948 and named Renold Chains Canada Ltd. (now Renold Canada Ltd.), for which a new head office was opened in Montreal in 1965. The Brantford works is a useful auxiliary of the Canadian business, meeting a demand that could be less easily supplied from the United Kingdom.

The Group's growth in Europe came by acquisition and expansion in both France and West Germany. Ever since the 1930s and the Renold-Coventry merger, the French manufacturing subsidiary had played a substantial part in Continental production and sales. The French-incorporated and

Renold-owned SA des Etablissements Brampton had a well-equipped factory at Calais. Repairs to damage during the German occupation and much additional building had made it a modern, busy and, taking one year with another, successful manufacturing unit. To this company was added in 1963 the business of Manufacture Lilloise de Chaines SA of Lille, makers of a variety of roller and conveyor chains, wheels and pinions. It now ranks as a manufacturing subsidiary of the Calais company. Group production in France was further rationalized in June 1965, when a new (Renold) Brampton wheel-manufacturing plant was opened at Charleville-Mézières in the French Ardennes, with buildings and plant on a site of 10½ acres, providing room for expansion. Wheel production, until then carried out at Calais, was transferred to Mézières, and the space thus freed at the older works was turned over to chain production.

While a wider framework was being provided for the French subsidiary, the Group took a step forward in West Germany. In June 1963, after discussions with the principals, it acquired the business of Arnold and Stolzenberg GmbH at Juliusmühle, near Einbeck, Lower Saxony. The German company was ready for a closer relationship. It had connections with Renold, partly dating back to a time before the First World War when Friedrich Stolzenberg represented the company in Berlin, and partly fostered by the settled Renold policy of tempering competitive effort with cordial relations on the personal plane.

Arnold and Stolzenberg is one of the leading manufacturers of chain in the Federal Republic, employing between 500 and 600 people. Its embodiment within the Renold Group afforded opportunities for technical co-operation, an extension of the product range and the ensuing advantages in market development, especially in the light of possible British entry into the Common Market. Arnold and Stolzenberg continue to operate as an autonomous subsidiary company; the change has been accomplished in harmony with the operations of the selling subsidiary, Renold Ketten GmbH, with its head office in Düsseldorf. The acquisition of the Einbeck company brought with it a 26 per cent interest in another German firm, Weydemann and Arnold, of Heidelberg, making chain wheels and pinions, and a 50 per cent holding in a firm making roller chain 7,000 miles away, an Argentine concern named Fabrica Argentina de Cadenas SA (FACSA). The principal object sought by the main transaction, however, was to provide for the further rationalization of marketing and the extension of chain-drive usage in Germany, where manufacture is scattered among approximately a score of companies, a third of which are about the size of Arnold and Stolzenberg and the others smaller.

Einbeck, over 700 years old, emerged unscathed from the Second World War. Near the town, at Juliusmühle, originally one of the many mills driven by the water power of the Ilme, is the chain manufacturing plant. The company was founded in 1909 by Otto Arnold and Friedrich Stolzenberg, who had in 1891 established a gear manufacturing concern named Friedrich Stolzenberg and Co. in Berlin. After Stolzenberg retired, Arnold took over his shares, and became sole proprietor of both enterprises.

Arnold came from an old family of Magdeburg industrialists. He made himself a name in industry while employed by Schaffer and Budenberg GmbH in Magdeburg-Buckau, who had numerous overseas establishments, and was the last chairman of that company before it passed into public ownership after the war. He died on April 26, 1945.

The firm's skilled workers at Juliusmühle have always been recruited locally, and are closely identified in spirit with the company. New homes have been built for many of them and those of other grades, and a housing estate has grown up nearby. Manufacture was resumed soon after the war; the firm was one of the first to receive permission from the British military authorities to reopen as chain makers. Up to the time of West German currency reform in 1948, many difficulties had to be overcome to obtain materials, but once economic conditions became stable, a systematic development and extension of activities was begun by the enterprise of the managing partners, Frau Thea Weydemann, a daughter of Otto Arnold and a woman of outstanding energy and force of character, and her husband, Herr Bernhard Weydemann. The overseas connections were re-established, and the export business developed anew as a result. With improving trade, research and development was intensified, and test rooms and laboratories were added to the works. The knowledge gained was made available to the industry generally. In 1959, the jubilee year of the company, Herr Bernhard Weydemann, as successor to Otto Arnold, further extended the company's interests and increased its share of the German home and overseas market for chain. This addition to the Renold Group was significant because it was the first time that Renold had acquired an entirely independent and fairly large-scale foreign manufacturing company having no element of British control.

During the years covered by the foregoing events, the Group moved into Austria with a selling subsidiary (Industrie- und Motorketten GmbH), increased the scope of Renold Chains Continental Ltd. (incorporated in Britain and controlling the selling activities in Belgium and the Netherlands), and acquired a 90 per cent holding in the Danish agency, Renold Grew A/S, while a majority holding in Cadenas de Precision S/A, of Barcelona, brought it into closer touch with markets which are likely to grow bigger if the modernization of Spanish industry continues.

The outstanding events of the period, however, were the absorption of the two English companies, Perry and Co. (Holdings) Ltd., Birmingham, and John Holroyd and Co. Ltd., of Milnrow and Rochdale, Lancashire. Their history has been outlined in Chapters 2 and 3. The merger of Renold and Holroyd in 1964 was without question the greatest event in Renold history since that of the Renold and Coventry companies thirty-four years previously.

By 1966, therefore, the Renold Group, already the world's largest organization for making and distributing precision chain drives and wheels, had become a rationalized organic whole, making and marketing a range of specialized technical products. Chain drives, and accessories to aid more efficient driving and fuller exploitation of the chain's versatility, remained fundamental to the Group, but with the Renold-Holroyd merger, two powerful forces were united in such a way that the strength each enjoyed as a separate entity was more than doubled by their union. In addition, the rationalization already begun before the merger had brought about a more convenient and logical deployment of the Group's manufacturing resources, such as the concentration of the conveyor chain part of the business at Burton, the disposal of all Perry interests (including pens and cycle fittings) other than chains, the reorganization of the Perry Works at Birmingham, and the introduction of a fully-mechanized flowline, incorporating many elements of automation, at the Coventry Chain Works.

The two chief factors dictating the changes at Coventry were the growing demand for automobile camshaft (0·375-in. pitch) chain and the problem of recruiting sufficient labour in the neighbourhood. The former is a result of the rising world output of motor cars; the second is related to this, since Coventry is a motor car-producing city, where competition for the available labour supply is intense. Originally, a cycle-chain flowline had been completed in 1955 and 1956. In the latter year, space at the works was cleared for the powder metallurgy department and for SCD adjuster manufacture when certain chain production was transferred to the works at Burton, but these changes did not touch the camshaft chain problem. This was solved by transferring the manufacture of much of the cycle and moped chain from Coventry to Perry Works, Birmingham, and rearranging the machinery in the north block at Coventry on the longer axis of the building. In this way the block was divided into two, one half for camshaft chain production on an expanded scale. The other half was allocated to the production of American Standard and a certain amount of cycle chain, the south block having been reorganized for the production of SCD adjusters and a number of miscellaneous products, and to accommodate the powder metallurgy plant; in the last-named most of the compacting presses were made by Manesty Machines Ltd., the tablet machinery-making subsidiary of John Holroyd and Co. Ltd.

At the time of writing, home and export sales are powerfully affected by the actions that the British Government has felt it necessary to take in the hope of restoring the United Kingdom's unbalanced economy. As a result, a correspondingly heavier responsibility devolves upon the overseas units of the Group. A factor in their strength is the Renold policy that, subject only to the maintenance of technical quality, stocks must be constantly and plentifully available for the customer.

A typical example of the stocking policy is provided by the conveyor chain factory at Burton-upon-Trent. In the early days of the project, the decision was taken to increase the number of conveyor chains immediately available ex-stock. Following a special sales analysis, the stock was therefore increased from approximately fifteen categories to more than sixty. This had a remarkable effect on sales and greatly altered the percentage of stock chain sales, which have now risen from between 10 and 15 per cent to a current figure of rather over 50 per cent. The result is of considerable advantage by reason of longer production runs and quicker deliveries.

The export drive, at any rate, did not find the Group wanting in the financial year 1965–6. Exports accounted for 40 per cent in value of all the company's chain sales, and an advance in the North American market of 29 per cent over the preceding year. Exports of chain to West Germany nearly doubled with an increase of 87 per cent. In the following year the Queen's Award to Industry set the seal upon the company's export performance—the reason for its conferment upon Renold Ltd. The Award was shared with John Holroyd and Co. Ltd., their part of the citation being for technical innovation, of which fuller details have appeared in Chapter 3. The Group was thus the recipient of a noteworthy double honour.

To sum up the export achievement: In total, something like 75 per cent of the Renold production is exported. Direct export sales of chains, by value, as has been stated, account for 40 per cent, but it is estimated that another 35 per cent is indirectly exported as components in complete British machines and plant. This is the outcome of a sustained sales effort in the face of unrelenting competition in the highly industrialized sectors of overseas markets. Despite that competition, sales have steadily improved in the traditional markets of Europe, Southern

Africa, Canada and Australasia; more recently, important contracts have been won in Japan, Germany and Italy for the supply of camshaft chains and adjusters; industrial business has expanded in the USA and in Latin America, and Renold products are available from 400 distribution centres in more than 100 countries—a sales and service network that undoubtedly had much to do with a performance of sufficient note to deserve the Award.

CHAPTER 6

RENOLD LIMITED

From comparatively early days in Renold history, the executive chiefs of the business have studied it as an organic whole, and have kept its articulation under constant review. They have sought to give it cohesion combined with adaptability suiting it to its various stages of growth. A dispassionate eye has been kept on the company and the Group as they have expanded; there has, as it were, been a built-in application of operational research. In an individual sense this attitude was characteristic of the founder. It prevailed among his colleagues during his later years, from the time when Sir Charles Renold's ideas and influence first became powerful.

CONSTITUTIONAL MANAGEMENT

A main component of the system is constitutional management as the foundation of the personal relationships of all who work in the Group. This approach, originating with Hans Renold, who was ahead of his time in this field of thought and action as in others and in many ways much more a man of the mid-twentieth century than of the late nineteenth, was closely linked with the policy imposed by him and followed ever since, 'to refuse to make other than the best, to have no second grade, and to maintain a fair price'. Sir Charles Renold fully shared his father's outlook, and in his turn elaborated and extended the measures to which Hans Renold's teaching led, with the success that makes their application possible in 1968 in a manner suited to the working lives of a community now numbering nearly 12,000 persons.

The principles underlying constitutional management were codified, and have needed little amendment since. They are to be found in Sir Charles Renold's book, *Joint Consultation over Thirty Years*, a second edition of which he was preparing at the time of his death. In the eighteen years since the book was published, the basic tenets of the 'constitution' have remained valid; the chronicle between 1956 and now is therefore devoid of great developments in this respect. The changes have really been adaptations made necessary by the growth of the organization during that time. To this general tenor of events there were two exceptions. One was the creation of the executive manager grade, about which more will be said below. The other was a company decision made in 1963 to accept the principle of willingness to join a trade union as a condition of employment in the United Kingdom works of Renold. This presented an aspect of a controversial subject often discussed. So far as Renold are concerned, the willingness to be a trade unionist on the part of a candidate for inclusion on their payroll is, within the framework of constitutional management, a logical outcome of a management philosophy which has always considered trade unionism to be a necessary component of a sound industrial structure, alike on grounds of public policy and in the interests of employer and employed. The recognition of union membership in the Renold case is in harmony also with the management's long-standing acceptance of the shop stewards as partners in the daily conduct of works affairs, regularly meeting to settle their problems.

Like constitutional management, the other elements in the system, while aimed at the achievement of orderly procedures, are of course directed to the performance of the actual work, regarded from the administrative, clerical and engineering points of view. They are intended to provide an efficiently articulated and, as far as possible, standardized system, or set of systems. They therefore include detailed procedures for decision-making, for maintaining the chain of command, and meeting the accompanying need for satisfactory methods of reportage and communication, the dissemination of information, and the presentation of accounts. In so arranging matters, care has also to be taken that flesh and blood are not overwhelmed by bureaucracy.

The immediate responsibility for the successful executive operation of the Group rests upon the general management team. In 1966 this consisted of the Group Managing Director, the Finance Director, the Sales Director, and the Works Director. The Finance Director also had in his charge certain overseas subsidiaries, selling subsidiaries being supervised by the Sales Director. The four members of the general management team, together with the Managing Director of one large group-within-the-Group, the Holroyd companies, are still in the same sense at the head of affairs, but history has again been in the making. Consequently, as we shall see, the situation, which served well until the Renold-Holroyd merger, has been further modified so as to fit it for the larger scale of present-day operations.

The practice has been for the general management team to run the business on a conciliar basis, the working parties with their general or special remits being called 'councils', with the Group Managing Director's Council at the summit. The councils consist of the four 'top team' members assisted by senior managers and other officials. All the same, top management keeps a firm hold on the power to make ultimate decisions on a number of reserved matters other than those allocated by prescription to the men in charge of the various works establishments, subsidiaries and branches. There is, however, consultation at all levels, covering all categories of work and the performance of the corresponding specialist functions general to the Group and its components. Each council has a clearly-defined sphere; the decisions it may make and the orders it may give are all carefully laid down.

The councils may deal with the affairs of a 'function' (e.g. the Employment Council with matters affecting personnel), or with those of a particular operational unit at home or abroad. There is a certain sensitivity about the use of the word 'committee' inside Renold Ltd., because of the bad name of so many committees in the world at large for their alleged loquacity and indecision. The committee structure and method in the ordinary sense are therefore in disfavour, but the term has to be employed, in a strictly Renold sense, for sub-bodies of the councils set-up to discharge some specialized duty. In sum, therefore, a 'committee' is simply an offshoot of a Renold 'council', the latter being a consultative organ composed of confidants and advisers whose views may be heard until the threshold of decision is reached. At that point, the process of consultation ceases; decision and action are the responsibility, according to the level of their importance, of a senior person or of senior persons.

This pattern of organization at the summit of the Group is not new. It grew in the past with the expansion of Hans Renold Ltd., especially after the First World War. There was a further extension and elaboration when the merger with Coventry Chain took place in 1930. It was then that constitutional management, and the merchandizing, budgeting and accounting standards introduced by H. G. Jenkins,[1] underwent their first big test of suitability in circumstances bringing many new elements into the business. They passed it successfully, came unscathed through the Second World War and have only recently emerged in their existing form, markedly influenced by the merger with Holroyd.

THE OVERSEAS SUBSIDIARIES

In February 1960, the procedures governing relations with overseas subsidiaries were recast, coming into operation in their new form a month later. The new model caused no direct alteration in the entire Group structure; its influence in this respect was felt later, because it led a year or two afterwards to new managerial arrangements described in the next section.

The overseas subsidiaries had grown greatly in number and diversity, and the immediate purpose in 1960 was therefore simply to improve their co-ordination, among themselves and with the centre. To this end, an Overseas Subsidiaries Director was appointed, Mr J. P. Halpin, then General Administration Manager. In this as in his other work he was at that time

[1] See Chapter 8.

responsible to the Deputy Managing Director, Mr W. S. C. Tully. His new duty was to ensure the co-ordination of all subsidiaries abroad. In 1960 these were Renold Chains Continental Ltd. (a company which, although based in Britain, controlled the sales in Belgium and the Netherlands), the Brampton Works at Calais, Brampton-Renold SA (the French selling subsidiary in Paris), Renold Ketten GmbH (the German selling subsidiary in Düsseldorf), Catene di Precisione SpA (an Italian subsidiary in Milan, now no longer trading), together with Renold Chains Canada Ltd., a selling company, and the two Australian companies, Renold Chains (Australia) Pty. Ltd., which conducted the works at Benalla, Victoria (set up with the encouragement of the Australian Commonwealth Government as a step in the official policy of Australian industrial self-sufficiency), and the company's selling organization in Australia, Renold Chains (Sales) Australia Pty. Ltd. The Overseas Subsidiaries Director is charged with the duty of generally advising the boards of these companies to enable them to make policy decisions.

The Overseas Subsidiaries Committee was simultaneously set up to make central Sales Division and Works Division experience available to the managements outside the United Kingdom. The Overseas Subsidiaries Director was appointed Chairman of this committee, the other members being the Works Director (Mr E. E. Hempsall) and the Sales Director (Mr Rex Bate); other chief officers of the company were required to serve on the committee from time to time.

These steps were given practical effect by appointing members of the committee to the boards of the overseas companies, the Overseas Subsidiaries Director having a seat on each of them. The boards of those nearer the United Kingdom thenceforth also included the Works Director if they were manufacturing companies; the Sales Director was similarly a member of the boards of the selling companies.

The committee was to report to the Managing Director and through him seek such board approval as might be necessary. It had to put forward recommendations on the constitution of overseas boards, the appointment of auditors and on matters touching the shareholdings. Power was given it, subject to report to the main board of Renold Chains Ltd., to appoint chief executives, and it could approve trading budgets, interim accounts and local dividend recommendations.

The objectives thus sought are defined in a private paper by Mr Halpin, presented to a Renold conference in 1965. He points out that in a world group there must be three main channels for exercising control or achieving co-ordination. These are: (*a*) financial control of the investment and its progress, (*b*) technical control of the specification of the product, and (*c*) sales co-ordination between the product exported and the locally-made product, particularly where supply of a common product, e.g. camshaft timing chain, is made in different places under different marks for the same end-use.

> 'In addition,' he points out, 'we have the advantage in Renold subsidiaries of advice from the Works Division on all production matters, so that the experience of eighty years is available to manufacturing subsidiaries. This is, strictly speaking, advice, but it in effect goes further because no sensible chief executive of a subsidiary would lightly fail to take advantage of this deep well of experience.'

He goes on to emphasize a feature characteristic of the Group as a whole:

> 'On the financial side it is imperative that there should be a strong financial thread of authority. We have seen examples of overseas corporations overwhelmed by overseas problems because the communications system was too slow or too insensitive for world group requirements. But perhaps even more important than authority or control is the need in a world group for a common management "language" in which to communicate.'

Another feature of the links between the Group's home and overseas interests is uniform management accounting. Broadly speaking, one company can be compared with another. Uniform accounts also provide the basis for Group budgeting and planning, although the operations of all subsidiaries are not looked at in the same way. Each must be regarded in the light of the economic and other circumstances of the country where it operates.

On technical control, Mr Halpin states:

> 'With a product like ours it is essential to see that standards and specifications established are adhered to throughout the world, and here the technical link must be a strong one and extend not only to approval of the specification of chain wherever produced, but to exchanges between the technical service experience of the United Kingdom and the overseas manufacturing subsidiaries.'

Thirdly, on sales co-ordination:

> 'The sales co-ordination again is an essential element because we claim to give a world service, and we must be able to see that the range we offer, whether from home or overseas manufacture, provides for the requirements of every kind of customer in every condition encountered and on commercial terms which are not mutually incompatible.'

These three points, and the passages chosen here to illustrate them, show that although, as has been stated, the 1960 rearrangements meant no radical change in the entire Group structure, they bore closely on the improvement in application of certain basic principles held by Renold general management.

In the same paper, Mr Halpin had something to say on a topic relevant at the present time. As it concerns the Common Market, and the Common Market concerns some of the more important of the overseas subsidiaries, it may usefully be included here:

> 'In the operation of a world group the assessment of the political situation is of paramount importance,' stated Mr Halpin. 'Thus, when the possibility of Britain entering the Common Market was being discussed [on the earlier occasion, in 1963] we in Renold were able to take a detached view. If we went into the Common Market, then we had a

foothold in France (and we subsequently acquired another in Germany), and were satisfied that we could organize our activities on that basis. When entry into the Common Market was refused, we were still satisfied that between our two footholds and our exports from Britain we could operate satisfactorily. This is still the case. I still believe that there would be great advantages, both to ourselves in the UK and to Europe itself, if we did enter the Common Market. I say this not because of any immediate advantage to Renold, but rather because this is the only basis upon which I believe that European trade can be substantially expanded and the only basis upon which sufficiently large markets can be made available to reduce unit costs by more efficient manufacture and distribution.'

Experience of the new method of co-ordination led to a modification in 1964, again reflecting the increase in Group activity during the intervening three-and-a-half years. By then it was necessary to assemble the companies in three groups. The first comprised the bigger integrated subsidiaries. These had been enlarged in number by the acquisition of the German manufacturing company, Arnold and Stolzenberg GmbH (with smaller subsidiaries of its own, FACSA and Weydemann and Arnold, which have been mentioned in Chapter 5, and a holding in a Belgian company, Mak et Coeurderoi sprl, which sold their products in Belgium), the French Brampton manufacturing and selling companies (which had acquired in the meantime Manufacture Lilloise de Chaines SA) and those companies manufacturing and selling in Australia and Canada. The second comprised the selling organizations in Europe and in America, including Renold Austria, Renold Chains Continental, Renold Ketten, and the American selling subsidiary, Renold-Perry Inc., although Mr Halpin in his capacity as Finance Director to Renold Chains Ltd. exercised financial supervision over both groups. Special arrangements were made for the Spanish company, Cadenas de Precision SA, Barcelona, in which Renold was then a minority shareholder, although now holding 61 per cent of the share capital.

THE EXECUTIVE MANAGER GRADE

With the acquisition of Arnold and Stolzenberg, Manufacture Lilloise de Chaines, other expansion in France, the parent company's participation in the companies in Argentina, Germany and Spain, and the establishment of more selling subsidiaries, for instance in Australia, all of which occurred in 1963, the time had arrived to consider further strengthening top management, giving the general management team more opportunity to delegate responsibility, and providing a bigger corps of senior executives in readiness for the still greater growth already foreseen.

Three distinct types of overseas companies are discernible: the major subsidiaries, manufacturing and selling in their own countries and probably exporting; they involved considerable Renold investment and were conducted by first-class men upon whom had been conferred a large measure of autonomy. Next came the selling subsidiaries operating as separate companies. Finally, there were the associated companies in which Renold had a minority investment. The needs of Renold management in these three categories or, in the case of the third group, the company's financial interest in it, were met by the measures outlined at the end of the preceding section, but taken together with the business of the whole Group they placed too heavy a burden of detailed work on the shoulders of the four chiefs principally concerned.

Concurrently, in Britain, Perry Engineering Ltd. was being transformed from a home subsidiary into a Renold works establishment, thus increasing the number of establishments within the works organization to five—too many for direct control by the Works Director alone.

In consequence, it was necessary to arrange effective assistance and provide reserve strength, a new senior executive 'support line'. To achieve this the board approved the upgrading of certain existing executive posts, giving the holders a substantially enhanced status, and the creation of several new appointments at the same level.

The various responsibilities, distributed in the new form,

were divided under the headings of finance, overseas, sales and works.

Under finance, Mr Halpin as Finance Director was given the assistance, at the new level, of a Finance Manager Home and a Finance Manager Overseas, the latter assisting the Finance Director in the exercise of his financial responsibility in regard to overseas companies whether integrated or not.

In the overseas field, Mr Halpin as Overseas Subsidiaries Director continued to be responsible for the major integrated subsidiaries. The day-to-day operation of these companies was in the hands of highly qualified chief executives, and it was decided that he should deal with them directly.

As to sales, the Sales Director, Mr Bate, continued to bear responsibility for the overseas selling subsidiaries and branches. Under him, again at the executive manager level, were a Sales Manager Home and a Sales Manager Export. The former operated the system as before, while the Sales Manager Export, through whom the additional overseas responsibilities were to be discharged, had under him a Continental Sales Manager and an Overseas Sales Manager.

As regards works, two Manufacturing Managers were appointed in the new grade under the Works Director. One looked after Renold Works, Manchester, and Burton-upon-Trent, and attended to overseas works liaison, and the other had the same supervisory role for the Coventry, Birmingham and Cardiff establishments.

The redeployment of senior management duties and the creation of a new grade of executive manager brought into the inner circle of top management a number of the company's leading men in the prime of life and of proven ability.

The Group Managing Director, Mr L. J. Tolley, believed that the rearrangements created the necessary executive strength for the immediate future and encouraged younger executives in the long term. In making them, he had also in mind the needs likely to arise from the movement into wider areas of positive power transmission which would occur following the merger with the Holroyd companies, then imminent. The acquisition of Crofts Engineers (Holdings) Ltd. has since made certain adjustments necessary, but the principle remains the same.

The executive manager grade has proved itself in the ensuing four years; it has realized the objectives sought in its creation, by relieving the Group Managing Director and the other general management team members of accumulating pressure as the Group has grown, becoming the 'support line' for the future and providing an incentive for able and ambitious men with general management potential.

RENOLD LTD.: THE NEW GROUP ORGANIZATION

The measures so far described, although timely and effective, soon needed further development when the Holroyd merger brought into Renold another group of companies.

In February 1967, notice was given of an extraordinary general meeting of the company, at which a resolution was submitted to the shareholders in the following terms: 'That the name of the company be changed to "Renold Limited".' The board explained that they had been mainly

> 'influenced by the fact that there has been a change in the primary activities of the Group from the manufacturing of precision chains and allied products to the establishment of a power transmission product organization. This change has been emphasized by the introduction of gears and other products which are manufactured by the Holroyd group of companies which became subsidiaries in 1964. In the light of this it is considered inappropriate for the parent company's name to continue in its present form. Steps have therefore been taken to obtain the preliminary approval of the Board of Trade to the proposed change of name, and, subject to the passing of the Special Resolution, it is intended that the change shall become effective as from the commencement of the company's next financial year on April 3, 1967.'

The meeting was held on March 15th at Renold House, and

the resolution was passed. On April 3rd a new Group organization took effect corresponding to the needs arising from the circumstances outlined in the letter. The entire business was rearranged in four divisional formations, the Chain, Sales and Overseas Divisions of Renold, and the Holroyd companies. Mr Tolley then became Group Managing Director, Mr Halpin being empowered to act as his Deputy Group Managing Director should Mr Tolley be absent. Mr Halpin retained his positions as Finance Director and Overseas Subsidiaries Director, with the prefix 'Group' to the first title, thus defining formally the scope of his wide responsibilities in the new general management structure.

The Chain Division corresponded with the former 'Works' Division, with Mr Hempsall at its head as Divisional Managing Director. The word 'chain', however, embraces a great deal more than precision chain alone; it takes in the whole range of chain accessories, the Renold sprag clutch, other (chain) clutches and couplings, and special tools and adaptors for use with all these. It distinguishes them from the many Holroyd products—wormgears, castings, machine tools, rotor milling machines, and the rest. Although the primary connection between Renold and Holroyd products is provided by the chain in alliance with the wormgear, the diversity brought about by the merger is such that time will be needed before it is completely integrated with the Renold organization proper. The new scheme takes account of this, so that although the Holroyd companies are in fact a 'division' of Renold Ltd., they continue as a Holroyd grouping.[1] The Sales Division, with Mr Bate as its Divisional Managing Director, was little changed; its responsibilities involve sales of the Chain Division's products throughout the world, the selling subsidiaries overseas, and the field selling force handling the Holroyd gear products.

To each of the three Renold Divisions—Chain, Sales and Overseas—has been allocated a board of management, the Chairman of which is the Divisional Managing Director. The

[1] The Holroyd companies, divisionally speaking, are now part of the Holroyd and Crofts Division, created in 1968 after Renold Ltd. had acquired Crofts Engineers (Holdings) Ltd.

sales and overseas boards have, moreover, been further strengthened by the enlistment of Holroyd talent.

In these ways the forces controlling the Group activity have been logically arrayed to manage the vastly enlarged apparatus of manufacture, marketing and administration in a large-scale international technical product company with many subsidiaries.

THE COMPUTER SERVICES

The use of electronic computers for business purposes is still only about fifteen years old in Britain. It is neither fully developed nor completely understood. Moreover, as a computer is an expensive machine, its purchase by any company calls for much prior study, whether it be regarded as a capital investment or a means to get work done. There is sometimes a tendency for directors, faced with tiresome problems caused by the complexity of modern manufacturing and trading, to feel that this seemingly magical piece of equipment, once bought and installed, will give instant efficiency, economy and profit. This is a mistaken belief. The prerequisite is not the computer itself, but effective non-computerized office systems. The computer as such cannot bring order out of disorder; only if the totality of its input services and materials are efficiently collected and properly prepared can its speed and versatility be of use.

Renold first considered the suitability of computers for their own requirements during the late 1950s. The company successfully avoided the mistake of putting the cart before the horse, or the computer before the systems. Plenty of time was allotted to think things out. There was also thorough preparation of all staff members who would be involved, both to accustom them to the idea in general and to make clear the ways in which the machine would affect their work.

The company ultimately decided to approach Leo Computers Ltd. (a subsidiary of J. Lyons and Co. Ltd.; it later amalgamated with English Electric Ltd. to form the company named English Electric-Leo-Marconi Computers Ltd., now in turn forming part of International Computers Ltd.). The reason was that Leo Computers had the experience, gained from their connection with the sophisticated clerical procedures of Lyons, that fitted them to make a feasibility study on the introduction of a computer at Renold House. Leo presented their report in January 1959, by which time it was plain that a computer would not be an economical proposition for Renold Chains unless it could be operated centrally at Renold House and ultimately cope with nearly all the clerical procedures necessary for works, sales, accounting and administration. This is explained in a private Renold memorandum as follows:

> 'The reason is that a powerful and therefore expensive computer is necessary in order to deal with the operations even of the smallest establishment (because even in a small establishment the amount of basic data is very large and the complexity of the programme as great—and sometimes greater—than in a large establishment). But on the other hand the total number of operations is really quite small judged by computer standards. Accordingly, the alternative idea (which on the surface has some attraction) of providing each establishment with a small computer is both impracticable and uneconomic, and only if substantially all clerical procedures are included can the savings be made to offset the considerable costs of buying and maintaining the computer and manning it with programmers and operational staff. The antithesis of the Renold position arises for example in an insurance office where the data (e.g. of insurance policies and their renewal dates) is relatively simple but the number of such policies is vast. In this sort of case, and there are many similar examples such as banks, the value of a computer is easy to prove because the computer instantly displaces large numbers of clerical workers and does the calculating operation in a negligible period of time and to all intents and purposes eliminates human error. The problem for Renold is a harder one than this.'

The feasibility study satisfied the Renold board that it was worth going further. Research continued until the stage at which a practical test became necessary.

> 'It was already known that theoretically a computer could work but what was not certain was whether it could in practice deal with the problems that arose in our particular business', states the memorandum previously quoted. 'It seemed pretty clear that the sales procedures, invoicing, payrolls and costing (which are logical, repetitive operations based on known facts) could be successfully operated by a computer and the same was true of accounting operations, but in the realm of production procedures—particularly with relation to production planning—the problem was difficult. Accordingly, it was resolved that the right way to test the matter was to select the most difficult operation on the production side and conduct a pilot scheme to see whether it could be successful. The selected department was the heavy chain department and thus, in October 1960, the pilot test scheme was started.'

Heavy chain was the most difficult option because its manufacture involves many variables, often of a 'one-off' kind, or partaking of 'one-off'-ness, connected with design, size and weight arising from customers' special requirements.

The principal matters covered by the scheme were:

Material control;
Work in progress and finished stock control;
The explosion of the programme into parts and assembly requirements, including material, manning and machine hours requirements;
Costing data;
The maintenance and revision of permanent data relating to material, parts and products.

The works mood was initially sceptical, but this scepticism was welcomed by the computer unit and by management because, should the computer fail to convert the sceptics, the odds against its success in practice would be heavy. As events turned out, the weight of the evidence removed the doubts.

The operation of the programs was tested on the Leo service computer in London in May 1963. The results justified

Renold hopes, and an English Electric-Leo-Marconi Leo III magnetic tape computer was bought and installed at Renold House.

In this complicated, prolonged and vital operation, Renold Chains Ltd. benefited greatly from their long-standing system-mindedness. Their clerical systems in general had been constantly updated for nearly half a century, both in the office and the works. The creation of new systems was unnecessary, and the staff were used to systematic working. Hence, although the computer would in the end cause changes in accordance with the technical and procedural laws of its own operation, the right climate existed in which it could do so in an atmosphere of orderly co-operation. Even so, making all allowances for the system-mindedness of Renold Ltd., the responsibility was a heavy one, borne chiefly on the shoulders of Mr Halpin.

The objective before the computer services is now the integrated performance of all clerical work, but management recognizes that the full benefits will take some years to realize. As time passes the process of transforming former manual clerical procedures via systems analysis and programming into computer instructions and thence to print-out will go forward with gathering speed.

TRAINING, EDUCATION AND THE CAREER STRUCTURE

The business of training is given a pre-eminent position in Renold Ltd. In this context, training means not only the instruction of school-leavers joining the Group at the outset of working life, but training, as youth passes into manhood, for such higher responsibilities as individual merit may warrant,and the 'refreshers' needed by seniors to keep them fit for the tasks that come with promotion.

The Group shares with other large employers the difficulties brought about by an educational system which is still evolving; the increasing opportunities for university education, whatever may be their advantages nationally, are limiting the field of recruitment among young men of the type required for business and industrial administration, owing to the emphasis placed, in many respects justifiably, on science and technology. Nevertheless, the fundamental task of the Chief Personnel Officer is to provide for adequate and balanced staffing of the whole complex of activities, and this calls for the preservation of a careful equilibrium in developing different abilities and skills in the right numbers. Another kind of balance is needed also in preserving the proportion between newcomers from the universities and the non-university entrants. These problems are further complicated because the post-war 'bulge' has passed, and the scope for selection has diminished. In the Renold Group's British offices and works, however, the various schemes developed since the late 1950s seem at present to fill the needs of the different categories. Older men, in the works especially, will sometimes say that the youngsters coming along are not up to the old standards. The expression of such views is a cherished privilege of older men, although some of them will, if closely questioned, concede that there are fairly bright specimens among the young entry. At any rate, the progress in Renold manufacture and marketing during recent years may be cited as evidence that the customer is not, to put it moderately, any more dissatisfied than in times past with the end-products resulting from the work of a later and younger generation of Renold-trained men.

Training is continued at different stages in a Renold career. Groups of 'middle' management are brought together for series of training conferences, each lasting two days, during which they learn more about the Group's operations and study aspects of business management of the kind that can be taught other than by experience in the office and the market place.

The Group holds periodical conferences also for sales engineers in the United Kingdom, but training is considered to be of equal importance throughout the organization and takes account of the overseas staffs. As has been seen, a high degree of uniformity has been achieved in accountancy, technical control and sales procedures. It involves a constant interchange of instruction between the head office and the subsidiaries. The Group Finance Director and Overseas Subsidiaries Director,

Mr J. P. Halpin, puts the matter thus: 'The technical man,' he says, 'must see the results of his decision in another country; the overseas salesman must come home to see new developments and pass them on to his colleagues. The cross-fertilization is fairly intense and is very productive.'

Great importance attaches to another kind of Group conference—the international sales conference. Formerly, conferences of this kind brought together, at approximately three-year intervals, directors, senior employees and agents from every part of the world, but by 1963 the greater range of operations led to the introduction of regional sales conferences for different territories, interspersed with full Group conferences. The first regional sales conference was convened at Houston, Texas, in February 1964, and the second at Rome in March 1966. In September 1965, the first of the Group conferences assembled at Eastbourne, Sussex. Although these gatherings are in a minor sense convivial occasions, they are predominantly practical, regarded from the point of view either of the platform or the floor, and are emphatically working meetings at which policy is clarified and laid down and developments, technical and other, fully and confidentially explained. They may thus fairly be described as training at the highest levels.

External, that is to say non-Renold, courses of value are not neglected, and Group executives attend the newly-instituted schools of management education.

Given a sufficient endowment of natural aptitude the individual in the Group's service will find management an ally in his or her personal effort to make progress. An example is the annual review of each worker's position and merits, upon which everyone who figures on the Renold payroll can count. Provision is made also to assist the employee in meeting the emergencies of life and assure comfort in retirement in ways that supplement the statutory minima of each country in which the Group operates. These are matters that need not detain the reader here; profit-sharing, pension and life assurance schemes are not of yesterday in the Renold companies: they formed an organic part of the conception embodied in constitutional management many years before the time which now concerns us.

Nevertheless, a far-reaching development of this complex and thorough effort to bring out the full potentiality of Renold employees occurred in 1958, with the introduction, on the basis of a previous scheme, of the personnel development scheme. As it is operated today, it provides for a systematic personal assessment aimed at meeting the five following requirements:

(*a*) Notification of employees considered to be capable of further development, and, in particular, of accelerated development.

(*b*) Easier filling of day-to-day vacancies.

(*c*) Maintenance of employee morale.

(*d*) A basis upon which to advise the employee on his vocational problems and prospects.

(*e*) The provision of information upon which to formulate training schemes.

The scheme is, therefore, in a large degree a method of 'talent spotting', and each employee knows that from the Group Managing Director downwards there is a strong desire to seek out talent if it be latent or insufficiently known and appreciated. This knowledge brings to the factory or office the conviction, at least to the ambitious, that it is worth while setting one's sights at long ranges and that exceptional merit will receive exceptional recognition. Management, in fact, is on the look out for those capable of filling jobs with a higher status than that of their existing occupations and particularly for any who are themselves potentially suitable for managerial appointments.

The means to achieve the desired outcome of the personnel development scheme is the potentiality review. This is not an easy task, nor is it performed in a perfunctory manner; it is in fact governed by detailed and thorough directives embodying much experience. Obviously a few people of outstanding ability will 'select themselves'. The borderline cases are more difficult, and involve questions such as these:

Do the demands of the job give opportunity for the display of potentiality (i.e. *real* potentiality, not showmanship)?

Is the employee good at assimilating new ideas?

Can he put new ideas into practice?

Is he mature?

Is he ambitious?

What has he done to help himself?

Is he self-disciplined?

Is there any evidence that he can control others?

Is he good at developing (i.e. bringing out the best in) others?

It may be thought that these are commonplace questions, but to get accurate answers to them is far from simple in a large organization, and if they are to be treated seriously they demand considerable concentration to arrive at a correct judgment. In one environment the same man will do well, in another badly. Opportunity may not knock owing to the very nature of his work. The immediate superior may be understandably if wrongly reluctant to further the departure of a good and competent assistant, although were this to be so it would run contrary to the entire Renold doctrine, because one of the chief tests of a good supervisor or executive is his own capacity for developing the Group potentiality of those responsible to him. The concept of Group consciousness, of the view wider than that bounded by the department, is intended to, and generally does, prevail.

A solution of these problems therefore turns on a correct technique of assessment before any decision can be made on an employee's working life, transfer, or promotion. The potentiality review, organized by the Chief Personnel Officer, is consequently an undertaking approached with considerable care. In the course of it the Group Managing Director and the other executives devote much effort to the search for management potential, culminating in the point at which they spend a fixed portion of their time in thorough examination of the names brought before them, considering in detail each of the employees recommended, and assessing, on personal knowledge and information obtained by prior and searching enquiry, the individual's standing. The classifications used are 'high', 'good', 'arrived' (i.e. the employee's potentiality is just maturing), 'new outlet required' (to bring out latent or perhaps frustrated talent), with a residual 'unclassified'. They also examine the inevitable quota of those with limited potentiality, for whom nevertheless there may well be openings somewhat higher up the line; this part of the process does not concern candidates chosen from the managerial grades, since the men classified in these grades have, by definition, already transcended a 'limited' potentiality.

It is necessary to be clear that the review is not restricted to those who may be assessed as worthy of advancement only in the most senior Group appointments; at more restricted levels, each year and in approximately the same way, a similar review takes place of the potential talent in each Group establishment.

After both kinds of review the papers are not pigeon-holed; they are emphatically action documents, and effect is duly given to the assessments.

HOW CONTROL IS EXERCISED

The channels through which the control of Renold Ltd. is exercised can be seen schematically in the organization chart (Appendix II), showing the position after the introduction of the new Group organization in April 1967. The link between the 'legislature' (the board) and the 'executive' (the Group Managing Director, the Group Finance Director, the Chain, Sales, Holroyd and Crofts, and Overseas Divisional Managing Directors, together with their executive manager colleagues on the Divisional boards of management) is provided by the first-named of these powerful men, the Group Managing Director. He exercises a continual initiative and great authority. We may recapitulate, and remind ourselves that he may, and does, consult others, but he gives the ultimate 'yes' or 'no', implementing or restraining action in accordance with board policy in which he himself has had a strongly formative share. There is nothing unusual in this, but the outcome at the point of production, sales or administration depends on other factors also. These are the clear operational channels, the distinction between operational command and services or 'functions' assisting the 'commander' of each of the 'rational entities' constituted by the works establishments and the subsidiaries, and the provision of a communications network capable of instantly informing the Group Managing Director of any significant change in the position of affairs, and likewise of telling his executive colleagues and subordinates, at each level, of those that affect them and must be handled by them either through action taken on their own responsibility or by reference to higher authority.

All this is grounded on well-tried theory, but its success depends upon the practical application or sensitivity of touch.

Thus, the Group Finance Director's responsibility for financial affairs is based either upon his own judgment and authority or upon those of the Group Managing Director and himself. This situation of itself offers a decisive means of control, both economic and technical, throughout the Group. It is further facilitated by close co-operation among the Divisional Managing Directors and their chief assistants who are members of the Divisional boards of management.

Combined with financial control are standard periodical estimates of future sales, the preparation of periodical trading budgets providing for the necessary capacity, costs and overheads, and the regular review of accounts. Fulfilment of sales and production programmes is checked by means of what is termed the periodic activity statistics meeting, held monthly, at which management has before it formal documents showing the different product groups at each stage in the thirteen four-weekly periods into which the year's working is divided. The figures thus presented state the stock brought forward from the previous period, with any adjustments made necessary by changing circumstances, and the outputs are compared with the budgeted figures. The statistics also contain a breakdown into home and export trading, and the periodic record closes with the stock, as it was at the time of budgeting and as it actually is, and the stock changes that have taken place.

Since periodic activity statistics are reviewed monthly and quarterly, are common to the Group and are standardized throughout the Group, so far as possible internationally, it follows that at any time Group top management has precise and recognizable information of all that is happening. Both Group and Divisional figures are correlated, and action can be adjusted to the known requirements.

But the system can only be justified if it can be prevented from becoming an end in itself. It must serve a practical purpose in the market place. This it appears to do; an insight into its effect in the home market is given by the Sales Divisional Managing Director, Mr Bate, in an address to a Renold conference in 1965. He said:

> 'We do have competition, but nevertheless our predominant position is a result of providing the best possible product, supported by a technical and selling service without equal which, in turn, is backed by stock held in every part of the country. . . . It is on account of this service, both technical and stocking, that we can beat low-priced competition. In other words, the customer pays for the value of what he gets from Renold, not just for a piece of chain.'

Mr Bate adds that the same may be said of the Group's trade overseas; Belgium offers one instance among a number: not far short of half the Belgian market demand for precision chain is supplied by the Renold subsidiary—at higher prices than those charged by continental competitors—despite the country's land frontiers with neighbours possessing competing chain manufacturing plants. Much of this and similar successes depend upon the availability of stocks, and the availability is the result of the watch that can be kept through the signals given in the periodic activity statistics.

The springs of prosperity are in the works. Efficient organization, sound systems, able sales engineers, good direction—all would be meaningless if the products did not match the apparatus which finds them their outlets. The then Works Director, Mr Hempsall (presenting a paper at the same conference at which Mr Bate made the observations quoted above), recalled words inscribed on stained glass in Hans Renold's old home, Priestnall Hey, near Renold Works,

Manchester, and now the employees' social club. They read:

> 'An eye that winces at false work
> and loves the true.'

The Renold Group's engineering eye has always winced at false work, and, it may be added, has not only loved the true but welcomed the new, if the new could stand up to the same discerning scrutiny. There is an identical outlook in the Holroyd companies. The Group cannot afford to be complacent, but neither has it any reason to suppose that it is falling behind technologically. The following quotations from Mr Hempsall's paper, with which this chapter may fittingly conclude, present an excellent example of the way in which Renold systems work.

> 'Briefly,' said Mr Hempsall, 'an establishment [works] is autonomous in all matters with the exception of task, for which it receives from headquarters broad guidance in the form of a trading budget which it interprets in association with current demand, stocks, circumstances, etc., to provide its work load, or, to use Renold language, its programmes.
>
> 'We begin with the four-year plan.[1] It finally becomes a total estimate of demand for all products in detail for the oncoming financial year and in product group totals for a further three years.
>
> 'When the four-year plan is completed and agreed, it is extended into a trading budget which is a prediction of what the activity is likely to be during the oncoming financial year. It enables the company to examine its manufacturing facilities and compare them with the demands which are likely to be put upon it. It also enables income and expenditure to be reviewed and available finance to be planned.
>
> 'The four-year plan and trading budget are finalized at Renold House, by the central merchandise section and, when approved, form the basis of Works [now Chain] Division planning. It will be appreciated that the trading budget is completed three months before the beginning of the financial

[1] This is prepared annually, i.e. each year the company envisages the work for four years ahead.

year to which it refers. It is an estimate which may during its life be anything from six to eighteen months ahead of action . . . it is often remarkably accurate. It is confined to broad terms and covers product group totals only. When received at establishments, the establishment merchandise department sets about the business of interpreting it into programmes for production departments.

'The decisions with regard to programme levels which may change from period to period, even from month to month, are made by the executive manager concerned, based on information from the establishment merchandise department in consultation with the establishment manager and his production manager. The factors considered are historical demand data, current level of demand, the position of finished stock, work in progress, raw material, capacity and the labour situation. The historical data are obtained from records kept in the merchandise section.'

He went on to explain that executive managers in their turn receive direction from the Works Director (nowadays, that is, the Chain Divisional Managing Director) based on decisions made at the activity meeting, which takes place during the week following the end of a period and at which the Group Managing Director, Group Finance Director, Sales, Chain and Holroyd Managing Directors discuss performance, demand, stocks and any special circumstances which may affect them.

'Once programmes have been fixed,' Mr Hempsall continues, 'the establishment merchandise control can issue a programme to the works departments for manufacture. This is, in fact, issued in the form of a four-period programme which provides output requirements for four periods ahead. It should be made clear that the issue of this programme periodically does not preclude alterations and amendments at any time and, indeed, modifications take place almost daily in the light of current requirements. The programme is then "exploded" into components and forwarded, together with bills of material, to works planning offices for manufacture. On receipt of these programmes and bills of material, the machine loading and processing of intermediate operations is carried out by works planning offices who "extend" all the components into raw materials in order to define raw material requirements by quality and size so as to prepare the raw material delivery schedules. . . . No material may be issued to the production departments until approval by the merchandise department, who ensure that it is in fact required for the programme issued.'

The procedure described relates thus far to standard or stock products, and there are certain variations with 'bespoke' orders for products specially made to customers' requirements.

The programmes are subject to the normal hazards of industrial life. Raw material may be delayed, machinery may break down, sickness and many other problems may have their effect, labour may be difficult to obtain. The task of production management and senior supervision is to solve these problems and ensure that the programmes are carried out from day to day.

Each week performance against programme is reviewed at works meetings over which the establishment manager presides, so that problems encountered can be examined and corrective action initiated. Finally, when the period is completed the results are reviewed and compared with the programme at the establishment council meeting, the chair at which is taken by the Executive Manager responsible and the members of which are the senior superintendents or production managers of the establishment.

Mr Hempsall points out that the picture is incomplete without a reference to the service departments. Machines and services are maintained, tools are manufactured to extremely fine limits in the toolrooms, in order to maintain the uniformity of products which ensures that components made today will fit precisely products perhaps manufactured many years previously. Jigs, fixtures and special tools are also made for bespoke products and manufacturing time and completion date are carefully phased to meet material deliveries. The overriding object is that products shall be received by customers punctually and in good condition.

CHAPTER 7

SIR CHARLES RENOLD, 1883–1967

Hans Renold's eldest son by his first wife, Mary Susan, *née* Herford, was Charles Garonne Renold, born on October 29, 1883. Aided by his parentage, the boy might well have been destined for a conventionally progressive career in the Renold company. It is on the whole seldom that the sons of brilliant fathers reveal the same ability in their own lives; fewer still are those who in one direction or another surpass the achievements of their sires. However, examples exist, and Sir Charles Renold was one of them. From his earliest years his influence was strong in the company's affairs. It waxed as time passed and the powers of Hans declined with age.

The intellectual and physical constitution which Charles Renold inherited from his father were developed and intensified by advantages coming from a domestic prosperity that had been lacking in the senior's earlier years, and by a wider education. Thus, mental ability, nourished by classical and liberal studies and sound technical training, produced in him a natural bent for the management of affairs. He had much in common with Hans, but was afforded wider opportunities; these he was able to define and apply in the industrial climate of an age ready to examine social and economic change and enthusiastic for technical progress. Management approaches, practices and techniques gripped his attention, because of their intrinsic interest, and because they seemed to him to deserve the best that careful thought and action could give. This profoundly affected the orderly expansion of the Renold business and, ultimately, the evolution of the Renold Group as we know it today. The interests of his father's firm and its successors were necessarily his first care, but his sense of mission produced reflections and studies of general value to management as a social science.

Charles Renold benefited materially from, and enjoyed, his privileged position as Hans Renold's son and heir, but his inborn zest was never confined to enjoyment of the rights without attending to the duties. After his schooldays at Abbotsholme, he went to America, and graduated in engineering at Cornell University. The choice of a university in the USA probably arose from the impression that North America had made on Hans, whose interest had been aroused by the enterprise and energy of American engineers, particularly in machine tool design and manufacture. Charles formally joined the staff of Hans Renold Ltd. in 1905, then being twenty-two, and in 1906 became a director. In 1908 he had his first opportunity to make a personal contribution to the company's organization and management. At that time, although a member of the board, he was in fact if not in name private secretary, or in modern terminology 'personal assistant', to his father. Thus placed, he was able to introduce the first of a number of changes—a clearly-defined method for keeping better records of instructions to the departments, and the actions taken to implement them. This was the starting point in his lifelong preoccupation with management principles and their application and adaptation to changing circumstances, whether technical, commercial or administrative. 'Orderliness' was a word frequently on his lips. His interpretation of orderliness as a matter of effective communications has coloured the entire development of the Group because, in fact, it has been his intervention as a young man, and his subsequent careful review and amendment of the system that have permitted the responses made by the organization to changes in the degree and, indeed, the kind of its operations for sixty years.

The 1908 reorganization of instruction-passing on an orderly system was followed by the introduction of a personnel department—then called the 'employment department', a startling industrial innovation at the time. In the same year the concept was further widened when a beginning was made in holding 'management meetings', foreshadowing the conciliar arrangements now governing the detailed execution of board policies in the Renold Group.

The young director had made a useful start. He was certainly not likely to be a genial but ineffective 'crown prince', dwelling in the shadow of his father and content to do so. The greatest share in these initial successes as a manager came from his own thinking, but it must not be forgotten that, once convinced of the necessity for any change, Hans Renold gave his support. This was the position when, in 1912, Charles returned to America to study the work of F. W. Taylor, the American pioneer of modern factory production methods. Back at Renold Works, he persuaded his father to allow the introduction of time-and-motion study, an event then remarkable in the British engineering industry. By 1913, his activities had also given all who worked for Renold a thorough understanding of the firm's structure and the boundaries of authority within the company. And this was not mere paper work divorced from practical shop floor and drawing office knowledge. The fact that Charles Renold was a qualified engineer was apparent also; he had by 1911 taken out three patents affecting improvements in change-speed gearing and machine design.

From 1912 he was in effect the company's Works Director. Each year added to his responsibility, and his co-workers at all levels saw that his ideas were soundly based, the outcome of prolonged thought and practical study. This, together with his genuine interest, devoid of affectation, in what was done and in all who did it, and his admiration, shared with his father, for the good workman without regard to status in the industrial hierarchy, had won him full acceptance.

In the Great War of 1914–18 Renold Works was chiefly occupied with munition making—'small parts in vast quantities' as Sir Charles Renold once described it. This brought its own management problems, and he began to examine the whole subject of what he came to call 'constitutional management' in the company. The history of 'constitutional management' and its subsequent evolution is told in detail in his own book, *Joint Consultation over Thirty Years*, published in 1950.

For the remainder of his career he never lost sight of what he understood, and what is understood in the Renold Group today, by the expression 'constitutional management', which is, briefly, a rational and orderly basis for the communal relationship within the organization. In its absence, the mainspring of internal harmony might be broken; the most thoroughgoing devotion, either to customer service or to the extension of markets for Group products (the two related and fundamental needs) might, on the most favourable assessment, be reduced in scale and effect, and at the worst produce frustration and restrict operations.

Constitutional management secures a just balance between the interests of the company and those working in it at whatever level. It has led to ungrudging recognition of the trade unions' place in the working economy of the Group expressed through a readiness to deal openly and fairly with the unions' representatives, primarily in the persons of the shop stewards, subject only to agreement that they in turn shall deal openly and fairly with management. This equipoise has resulted in many years of co-operation, with only occasional failures to agree and no widespread nor prolonged conflict of views to spoil the record.

The Works Director *de facto* became the Works Director *de jure* in 1916. The policy of constitutional management was formally introduced in 1917, whereupon a shop stewards' committee was formed at Renold Works, which was then the company's only factory. Charles Renold had simultaneously in mind the wider field of organization in industry generally. Apart from the fact that Hans Renold Ltd. was likely to evolve into a much bigger business, he appreciated on the international scale the share industrialists would be expected to take in restoring the economy after the war. His views were strengthened in 1919 when he became Managing Director, and first discussed with his father the idea of combining the Renold, Coventry and Brampton companies. The preliminaries to the merger occupied much of his time for the next ten years. The merger itself finally came about in 1930, principally as a result of his diplomacy.

Nine years after the end of the First World War, Hans Renold, who had been at the head of affairs for forty-nine years, decided to give up the burden of the chairmanship, designating Charles Renold as his successor, with Charles's cousin R. O. Herford as Deputy Chairman, H. G. Jenkins[1] as Sales Director and Mr E. J. Parish, now retired, as General Works Manager. With these four there came into being the company's first clearly recognizable general management team drawn from a younger generation than that of Hans and his contemporaries in age; in previous years Hans Renold had in effect controlled the business personally, despite the considerable modifications and adaptations introduced by 'Mr Charles'.

The new Chairman's first period of office was short, because of the success of his own efforts in negotiating the merger with Coventry-Brampton in a single company which then became the Renold and Coventry Chain Co. Ltd. When the time came to give organic effect to the arrangement, he left the chair, and ceded the appointment in the new organization to Arthur Brampton, the leading figure among the Coventry and Brampton directors. By doing so he believed that the Coventry and Brampton forces would be more happily integrated in the enlarged company, since Arthur Brampton was not only a man of great business ability, but popular and trusted among the newcomers who, finding themselves in a world of entirely new industrial concepts, were anxious about their future. Charles Renold therefore yielded a substantial point, the concession of which went a long way to ensure the right spirit in the amalgamation. His effective leadership was not really curtailed; he became Managing Director as well as Deputy Chairman of the new board, and so remained until his father's death in 1943, when Arthur Brampton retired and Charles resumed his former place as Chairman. Thereafter he presided over the company as it has evolved successively into Renold Chains Ltd. and Renold Ltd. almost to the day of his death.

When he again became Chairman, the company was once more engaged in munitions production. For a year he held

[1] Herford died in 1955, and Jenkins in 1961. Further reference to them is made in the next chapter.

the posts of both Chairman and Managing Director until in 1944 Herford, released from a wartime engagement with the Ministry of Supply, took over as Managing Director. At 60, Charles Renold enjoyed the vigour of mind and body of a younger man, was no less ready to examine and test new ideas, and constantly sought to improve the organization of what had become a great international business.

The Second World War, like its predecessor, was rapidly and drastically influencing social and industrial change. When it ended, the problems of organization and management were more diverse than in 1919, partly because the war had been 'total', and partly because the company itself had much more to do in re-establishing its international branches and agencies, many of which had been fought over, bombed, or seized by the former enemy.

High on the British Government's list of necessities for the reconstruction of civilian life and trade was a thorough examination of the ways in which managers were to be found and trained for industry. In 1946 the Board of Trade therefore encouraged the establishment of the British Institute of Management, the purpose of which was to study and discuss the subject. The progressive reputation of the Renold and Coventry Chain Co. Ltd. and its Chairman were recognized when Charles Renold was asked to accept the chairmanship of the new institute. It was work he enjoyed and was confident that he had the experience to do well. One of the first of many contributions he made to the cause of the BIM was his case history of joint consultation as it had been practised in the Renold factories, set out in the book already mentioned. He held the BIM chairmanship until 1951, when he became its first Vice-President. His record was by then becoming more widely known than in his own company or in the north of England; in 1948 he was knighted for his services to the cause of good management and the development of humane and progressive ideals and practices in industry.

With the Renold-Holroyd merger, the physical size and international extent of the Group were sufficient to emphasize more clearly the need for decisions on managerial, administrative and communications procedures akin to those called for in the largest companies. Sir Charles Renold had reviewed requirements of that kind in a paper presented to the BIM during its winter session of 1948–9. Entitled 'The Organizational Structure of Large Undertakings; Management Problems', the document aroused widespread interest. It may fairly stand as an epitome of his philosophy on the subject.

After remarking upon the changes he had witnessed during his working life, especially the continuous expansion of the Renold enterprise, he examined the limitations placed upon the leadership of large undertakings—'leadership' in the sense of personal supervision of activity, 'the great width of the gulf that separates those at the top who formulate policy from the unit leader who has to apply and observe it. I suggest' (he continued) 'that it is the point at which this gulf becomes too great to be bridged by personal contacts that the peculiar problems of "large scale-ness" put in their appearance.'

He explains that, concurrently, other difficulties arise. For example, the channels of operational command must be kept clear. 'Large scale-ness' involves increasing numbers of specialists, whose necessary incorporation in the system must be reconciled with the maintenance of a clear channel for operational command and the delegation of operational authority. The specialism—whatever it may be, sales, personnel, accounts, and so forth—has its own chief within the organization, and the specialist tends to look to him for directives. This prompts the question whether the specialist responsible to his own higher functional authority is, in his relationship with the 'unit' or operational 'commander', an adviser or a 'commissar' to see that the commander toes the line. This indicates one of the ways in which 'large scale-ness' generates its own problems, which seem to fall under one of three headings. These are:

(*a*) The difficulty—arising from the wide separation between the operational unit and headquarters—of forming at headquarters true assessments of situations. Sound judgment needs both facts and 'feel'; facts can pass the filter screen of the hierarchy but the 'feel of the situation' cannot.

SIR THOMAS ROBSON

MR L. J. TOLLEY

(*b*) Effective leadership of the operational unit is liable to be undermined. This is due not so much to the multitude of overriding uniformities to which the unit commander is subject, as to the necessity for embodying these in regulations which can hardly be other than rigid and insensitive to local circumstances—again because of the wide separation between the operational unit and headquarters.

(*c*) The conflict between the two aspects of the role of the functional specialist—that of adviser to the operational commander at any given level and that of 'commissar' to see that he 'toes the party line' in respect of the specialism.

Having defined the difficulties, Sir Charles went on to examine the lines along which the bounds they set to the manageable size of an undertaking might be enlarged. For this purpose he sought to establish a basic 'rational entity' (the operational unit). The hallmarks of the perfect operational unit, he suggested, are four:

> That it should be a rational entity.
>
> That it cannot be sub-divided into smaller rational entities.
>
> That the relations within it can be treated as domestic to itself,
>
> and
>
> That it disposes, within itself, of the resources necessary to its day-to-day operation.

If these conditions are met, it is a unit suited to one-man command. To illustrate his point, he chose the analogy of a warship—essentially a rational entity, in which possibly competitive or influential specialisms (power supply, gunnery, catering and so on) are given their due weight and no more because, apart from the ship as a whole, none has meaning in itself. Thence he extrapolates from the ship to the squadron, from the squadron to the fleet and finally to the Admiralty Board which operates in terms of a number of fleets.

Applying the analogy to industry, he saw the origin of many organizational problems in the fact that industry's operational units are not always rational entities in the sense described.

Admitting that no rule can be laid down, and that it may seldom be possible to make an industrial unit conform exactly to the naval pattern, he nevertheless affirmed that there is room for closer conformity than exists, and that the advantages of this closer approximation are often sacrificed in pursuit of 'some specialized aspect of efficiency which may well be bought at too high a price.

> 'For instance, in a large undertaking that includes a number of manufacturing engineering works it may appear attractive to set up a central toolmaking establishment to serve a group of such works. Greater efficiency in toolmaking may thereby result, but the manager of each of the several works thereby loses control of one element necessary to the fulfilment of the tasks laid on him—he no longer disposes of all the resources necessary to his day-to-day operation. The overall effect on efficiency may well prove disappointing. Centralizing the purchasing for a large undertaking is another case in point. How often are the advantages and economies more than offset by frustration in the operational units?'

The problem of the functional specialists is solved in the ship because, while the 'how' of what they do is defined at a higher level, the 'what' and 'when' of their actions never by-pass the captain, although he in turn commands the ship's company within a framework of regulations emanating from his superiors. Sir Charles believed that the same conception and practice could and should be applied in industry.

He paid attention also to the divergent forms of organization open to large undertakings, e.g. a 'federal' structure, with a parent company which may or may not be a holding company only, surrounded by semi-autonomous satellite companies, as distinguished from a structure in which the branches are energized and controlled from a central source of authority, both technical and financial, within a commonly observed code of instruction and reportage served by a uniform communications network and a uniform system of administration. Renold practice has been to adopt the second method; its inherent economy and effectiveness commend it, and so does its relative simplicity, provided that it is continuously and carefully studied, never taken for granted, and always kept up to date. Signs of the 'federal' structure sometimes appear in the company's history, but usually only in the interim stages of a specific expansion involving another company or group of companies, before a merger or acquisition and pending the final integration in the Group structure and procedures.

Sir Charles Renold stated the results of his analysis in the following propositions:

(*a*) Sound judgment cannot be based on reported facts alone, but requires also the 'feel' of the situation.

(*b*) While the facts can survive a long chain of reportage, the 'feel' cannot. Hence the point at which judgment should be formed and action taken should be as near as possible to the point at which the situation has arisen, since at such a point both 'feel' and facts are accessible.

(*c*) The large-scale undertaking is, in this respect, faced with a serious problem in the great gap between the operational units, where the situations arise, and headquarters.

(*d*) The difficulties are twofold. The first is that of forming sound assessments of problems on account of the length of the lines of communication—the chains of command. The second is the complexity of the overriding uniformities which condition command at operational unit level, and the rigidity of their operation, which tend to undermine the reality of command at that level.

(*e*) The basic approach to a solution of these problems turns on the conception of the operational unit as a rational entity, and of the progressive groupings into larger and larger units of command, being themselves in turn also rational entities. To the extent that this conception can be applied, the points of judgment and of action are brought near to the situations in question, since each level deals with matters of the same order of generality and which are domestic to the particular level. Moreover, a rational entity lends itself naturally to effective command.

(*f*) Some assistance in simplifying relationships may be

available along the lines of a federal formation composed of associated satellite undertakings. By this device the top link of the chain of operational command may, in effect, be discarded and thereby permit simplified systems of relationships.

(*g*) Nevertheless, there still remains a baffling problem in the relations between functional services and operational command: the dilemma which he labelled 'commissar' or 'adviser'. The solution of this turns on the acceptance of a fundamental assumption, viz. that the operational commander at each level desires to conform and needs no forcing into conformity by 'commissars'.

(*h*) Finally, he acknowledged that while this basis of solution permits success, it does not, of itself, ensure it; he therefore suggested the need for an independent organization-inspection service to report on how the organization is, in fact, working.

He described the satisfaction he derived from teaching and practising management in an article on management as a career which appeared in *The Financial Times* of October 31, 1955:

'To one who has found such a career deeply satisfying it is difficult to know where to begin. The most obvious reward is the financial one and it is almost axiomatic that the big salaries, apart from a few star performers among the technical specialists, go to management, and that they rise steeply up the hierarchy to top management. This is the natural corollary of the facts: first that, among the factors which determine the success of an enterprise, the quality of its management is the most important, and, secondly, that managerial ability, despite all that can be done by education and training, is in short supply.

'But the satisfactions go much deeper. Management is power, if not *de jure*, certainly *de facto*. Power can, of course, be abused and is not then a legitimate source of satisfaction. But power is also synonymous with opportunity for service.

'Management serves directly three interests—those of the shareholders or owners, those of the customers or consumers and those of the employees of the concern. To a greater or lesser degree the satisfaction and well-being of each group is in the care of management. This is the case in a very direct and intimate sense for the employees: their very way of life is coloured by the kind of management that leads the enterprise. Their possibilities of satisfaction derive not only from the actual conditions of work and employment but even more from the extent to which the management succeeds in creating a sense of participation in a common purpose. Such leadership carries its own satisfactions.

'But beyond these immediate and domestic fields of service and the satisfactions derivable therefrom, there are wider fields. For an industrial country such as ours the well-being of the whole community is dependent on the successful conduct of its industry, which in the last resort means the successful conduct of all the individual concerns that together make up industry. As the well-being of each of these units is, in the main, determined by the quality of its management it is obvious that management, as against all other factors, has the greatest potential contribution to make to national well-being. It would be hard to imagine a more inspiring justification for any career.'

In 1956 Sir Charles Renold celebrated his seventy-second birthday, at which time he had completed over fifty years with Renold Chains Ltd. and its forerunners. Eight years later, at eighty, he was still in unabated vigour of mind. The dinner which marked the anniversary was made noteworthy by the combination of powers plainly manifest in the youthful octogenarian whom his colleagues and friends had assembled to honour. Dr B. V. Bowden (now Lord Bowden of Chesterfield), the Principal of Manchester College of Science and Technology, Mr H. W. Swift, a retired General Manager of the company, and Mr L. J. Tolley, the Managing Director, acclaimed not only Sir Charles Renold's directorial and managerial achievements with Renold Chains Ltd., but emphasized the impressive record of his public work, especially in the cause of higher education.

Sir Charles's educational work had indeed extended progressively with the years, largely on lines connecting it with his ideal of the educated managerial technician. Although a serious and painstaking magistrate (he was appointed a Justice for the City of Manchester in 1945), it was in regions harmonizing with his own main preoccupation that he gave his principal public service. From 1925 onwards, he was closely associated with Manchester College of Science and Technology (now the University of Manchester Institute of Science and Technology). In that year he became a member of the Manchester Education Committee's sub-committee for the MCT. From its formation in 1956 he was Chairman of the MCT Planning and Development Committee until July, 1965, when he became Deputy Chairman. On his eightieth birthday he was a Vice-President and Deputy Chairman of the College Council. In his person, he stood also for a company which, ever since Hans Renold's day, had been one of the college's most generous benefactors. In 1962 he declared open the new lecture room block, named in his honour the Renold Building.

A further commemoration of their Chairman's eightieth year was his company's endowment at the college of a post-graduate research Fellowship in mechanical engineering. Known as the Sir Charles Renold Fellowship, it is renewable annually for three years at £800 a year, the holder undertaking a course leading either to the degree of M.Sc. Tech. or Ph.D.

Accompanying Sir Charles's service to the college was his membership, beginning in 1936, of the Court of the University, and from 1938 of the University Council also, and his chairmanship of the University Buildings Committee from its establishment in 1946 until 1962. This was no sinecure. When he relinquished the appointment, the council of the university recalled in a resolution of thanks that:

> 'Sir Charles was appointed chairman of the committee on its establishment in 1946 and a task awaited him more difficult, more complicated and more time-consuming than anyone could then have contemplated. Preliminary thinking on the future of the university had taken place throughout the war. That thinking, inevitably tentative in character, had

been based on the assumption that the university would expand. None could see how dramatic and rapid that expansion would be. A long-term planning problem was involved, and after the collection of basic data and the employment of consultants a solution reached. A master plan was adopted to which Sir Charles' contribution was paramount, and has not in any serious detail been departed from. The uniqueness of this achievement cannot be overemphasized. To evolve and to bring steadily towards realization a plan under conditions which remain stable is one thing; to do the same when every factor has, under the pressure of events, become a variable, is quite another. It is surely not the least of Sir Charles's contributions to the present and future face of this university that, meeting every change, coping with every pressure, he and his committee have kept faithfully to the original design.'

Lifelong absorption in management problems prevented him from passing over any detail of possible relevance, even if apparently remote from the conduct of business in the 1960s. His attention was drawn in 1966 to the observations on Army organization and its relationship to organization in general made many years ago by General Sir Ian Hamilton, the British Commander-in-Chief in Gallipoli during the 1914–18 war, in his book *The Soul and Body of an Army*. Sir Charles at once obtained a copy; commenting upon some passages in it, he wrote to the present author:

'I got hold of a copy of *The Soul and Body of an Army* and have now finished it, having had plenty of leisure last week.

'I found it quite fascinating—both as bearing on my problem of the wherein-ness of morale and as taking me back to a world in which I was a young man and which is now quite gone—the world of the turn of the century.

'You are quite right about the Army being more complicated than the Navy! But for that very reason it seems to me easier to grasp the significance of the basic unit of organization being itself a "rational entity" by using the ship, squadron and fleet as the illustration.

'Then again—Sir I. H. argues that the basic unit—rational entity—for an army should be a *Division*—about 6,000 men—of the Roman Legion. That is too large a basic unit for industry. But all very thought-provoking and therefore exciting.'

He was twice married. His first wife was Margaret Hilda Hunter, of Manchester, whom he married in 1909, and by whom he had three sons and a daughter. She died in 1958. Two years later he married Noël Garry, of Sale, Cheshire, and Dublin, but in 1966 she, too, died after a long illness. The blow was heavy, perhaps the heaviest he had borne in a long life. Even so, his personal sorrow was masked, except among those who knew him best, by his determination not to let its effect impair the performance either of his duty at the head of the Group or as host to the many people whom he necessarily met in the course of the day's work. The second Lady Renold was a devoted and practising Catholic, and her husband was deeply impressed and affected by the fortitude, based on her faith, that she displayed in her last illness. He spoke of it at some length to the present writer, remarking how much he wished he could accept Noël Renold's own beliefs himself. But this was not possible for him in conscience. 'I think,' he said, with the slightly mischievous smile that often accompanied his very lively wit, 'I should nevertheless like to look upon myself as an associate member.' But, courageous as he was in his grievous deprivation, the bereavement, added to his own increasing physical disability, hung heavily over his last months.

When, following the annual general meeting of Renold Ltd. on August 3, 1967, Sir Charles signified his wish to retire from the chairmanship, his link with the Group's mighty past and its future development remained unbroken. At the wish of his colleagues, he consented to continue in office as a director, and to accept the appointment, unique up to that time in Renold history, of Honorary President of the company, a recognition of the debt of honour due to his special position and extraordinary career at the head of affairs.

Death came peacefully to him at last on September 7, 1967,

in St Joseph's Hospital, Whalley Range, Manchester, where his second wife had died. The funeral was a strictly private family occasion, but at the memorial service in Manchester Cathedral on September 20th, representatives of the great city, its university, and the far-flung Group he had done so much to organize and articulate, assembled to do honour to one of the north's outstanding men of affairs.

In one respect he outdistanced his father. His negotiating ability, and the industrial diplomacy that had served so well in the Renold-Coventry-Brampton merger of 1930, found expression again in the discussions leading to the junction of the Renold and Holroyd forces thirty-four years later. In the earlier event, Hans Renold left the main business to his son, whereas the latter, at a comparably advanced age, was an active principal up to and including the formation of Renold Ltd. He was also the man who bridged the years between the original emergence of 'large scale-ness' in Renold and its latest phase. As the small 'top teams' of executive directors responsible for general management changed, the Chairman long remained the same, ready to examine new ideas but able to judge them with the insight developed by mature understanding of managerial techniques. He was therefore blessed above the generality of patriarchal figures, in being both loved for himself and welcomed as chief and in council by no less able but much younger colleagues.

To succeed him as Chairman, the Board appointed Sir Thomas Robson, one of several directors who have served the Group although neither by origin nor training Renold men. Sir Thomas, however, had long been intimately connected with Renold affairs, having taken a leading part in the formation of the Renold and Coventry Chain Co. Ltd. in 1930, as a member of the staff of Price Waterhouse and Co., then as now the company's accountants and auditors.

He was born at Newcastle-on-Tyne, and after taking a degree at Durham University served overseas during the First

World War in the Royal Artillery. In the course of his service he was appointed MBE.

Sir Thomas took the final examination of the Institute of Chartered Accountants in 1922, winning the W. B. Peat Gold Medal, and joined Price Waterhouse in 1923; he became a partner in 1934. In the ensuing years he has served as President of the Institute and as a member of many important Ministerial and other bodies.

He entered upon his appointment as Chairman of Renold enjoying the complete confidence of his co-directors, firmly founded upon his known devotion to the Group and his knowledge of its problems and potentialities.

CHAPTER 8

THE GENERAL MANAGEMENT TEAMS AND THE BOARD

The way in which the general management team's authority is exercised has been considered in more detail in Chapter 6. Here the purpose is briefly to trace the careers of the men who have held the key posts, or their past equivalents, since 1943, when the concept of the 'top' or general management team first evolved distinctly. The holders' titles and the division and extent of their responsibilities have varied from time to time in twenty-five years, and will undoubtedly continue to do so as new needs and circumstances arise, but under the guidance and supervision of the Chairman and the executive and administrative authority of successive managing directors, the principle of a division of responsibility among the executive directors for works, sales and finance has been the general pattern.

The men who took the first steps in the post-1945 reorganization of the company and its subsidiaries consisted primarily of Sir Charles Renold, R. O. Herford and H. G. Jenkins.

Herford, a cousin of Hans Renold, and originally one of his graduate apprentices, retired in 1953, after forty-four years' work. During the 1914–18 war he took charge of all the technical aspects of production for munition making, and subsequently of all manufacturing activity in Hans Renold Ltd. His dominating personality and high engineering talent give him a lasting place in Renold history. When he completed his years of executive service he remained a director. He died suddenly in 1955 while visiting the company's agents in the Argentine.

Almost immediately after Herford and Jenkins had taken up

their duties at the summit, they were joined by Mr Oscar Bertoya, Mr W. S. C. Tully and E. J. Parish; the last named, like Herford, began his Renold career as a graduate apprentice, and rose to be General Works Manager. He died in 1968.

Jenkins retired in 1948, after forty-three years with the company, having joined Hans Renold Ltd. in May 1914. He was the first general management team member to be officially described as Finance Director, a title he received when he became a director in 1919. A considerable part of the existing internal practice in the operation of standard routines affecting accounting, budgeting, and office and sales systems is based upon principles initiated by Jenkins between 1914 and 1924. In 1924 he became General Sales Manager, and after the 1930 merger of Renold and Coventry-Brampton played a leading part in product rationalization. Although reaching retirement age in 1948, he continued in full-time service for a few years more as Managing Director of Renold Chains Continental Ltd. He died, aged 75, in 1961.

Six men formed the second executive team—Mr Oscar Bertoya, Mr W. S. C. Tully, Mr W. V. Foley, Mr L. J. Tolley, Mr J. P. Halpin, and Mr H. W. Swift, the Deputy General Administration Manager. Mr Swift brought a financial and organizing brain of the first order to company affairs, and had much to do with the transition of the company from its nineteenth to its twentieth century structure.

When Herford retired from the Managing Directorship, the succession devolved upon a Managing Director, Mr Bertoya, and a Deputy Managing Director, Mr Tully. The intention was that the two should work closely together, sharing the field between them, Mr Tully concentrating particularly on the financial aspects. As senior partner in the combination, much of the initiative of the office naturally fell to Mr Bertoya.

Mr Bertoya, who was born in 1900, was trained in early manhood as an engineer, and entered Hans Renold Ltd. in 1923 on the export sales side. His command of most European languages reinforced business ability of a high order, and brought him much service overseas. Shortly after joining the company he was sent as manager to the Brussels office. Thence he went to Germany where, stationed in Berlin, he was at the head of Renold Ketten GmbH, the German selling subsidiary, from 1929 to 1935. In the latter year he was promoted European Territorial Sales Manager, holding at the same time the general managerships of a number of the European selling subsidiaries. When war broke out in 1939 he was in Britain. During the conflict he organized and managed a factory run by the company for the Government at Newton Mill, Hyde, Cheshire, where a staff and labour force of 500 produced precision tools for making small arms ammunition. When the war ended he was appointed Deputy General Manager of subsidiaries. In 1948 he was made General Sales Manager of the whole company, the position he held until his appointment as Managing Director in 1953. He retired in 1962.

Apart from his success in pioneering the European subsidiaries Mr Bertoya, looking back on his work, sees his chief contributions to the Group under five headings. First, when he was General Sales Manager, came his reactivation of the sales force and the European companies after the Second World War. Second, it fell to him to overcome the difficulties of the post-war austerity period, when steel was scarce and pent-up demand heavy. The third heading in Mr Bertoya's own assessment of his task was the creation of a cadre of new men for the Renold sales division and, subsequently, when Managing Director, for staffing the organization as a whole. Possibly—although this is the author's observation and not Mr Bertoya's—his greatest single triumph was the fourth in his list: the successful achievement of the series of expansion plans to meet the demand, long foreseen, for Renold products throughout the world and to ensure that, in his own words, the company should 'never again be short of capacity'. Fifth, standing close to the expansion plans, and no less significant in the long-term, was the establishment of large-scale facilities for research and development. Added to all these were other successes, the most important of which were the inauguration of the conveyor chain factory at Burton-upon-Trent, Staffordshire, the acquisition of the Perry enterprises, the establishment of the factory at Brantford, Ontario, the acquisition of the Anchor Chain Co. Ltd., Oldham, Lancashire, and the building of new branch offices in Britain and abroad.

Mr Bertoya has also become a notable figure in industry at large; he has been prominent in several national bodies concerned with the care and protection of British and north-country industrial and commercial interests. His vigorous oratory has brought him many requests to address these organizations, and his knack of forceful and picturesque analogy, to which he gave full rein in speeches at a number of Renold sales conferences, are vividly remembered by his hearers. Although he retired from the Managing Directorship in 1962, he remains a member of the board.

Mr Tully, the Deputy Managing Director, joined the company with the title of General Administration Manager in 1946. He became a director shortly afterwards. He possessed a profound knowledge of financial problems, based upon a successful business and official career in the former Indian Empire. A west-countryman, born in 1900, he became a chartered accountant in 1922 and thereafter spent twenty-three years in the East, where he rose to be senior partner in India of a large British firm of East India merchants and industrialists. In 1945–6 he was Director-General of Consumer Goods to the Government of Bengal, returning afterwards to England and joining Renold. In the company's post-war period his work was vital in strengthening the financial and administrative sides, and made a permanent mark in updating the company's planning and budgetary control. Of him, Sir Charles Renold wrote: 'It was realized that we were in process of becoming a world-wide "Group" of closely associated companies, and one of W. S. C. T.'s early tasks was to reshape the company accounts to give expression to this new conception. The elaboration of procedures of all kinds to implement the conception has been one of W. S. C. T.'s outstanding contributions.'

During Mr Tully's executive career there were four capital issues, for all of which he was responsible, among many other financial and administrative activities arising either as a result of the war or through Group and company expansion. As befitted an accountant, he was a stickler for exact information, and knew how to get it without causing resentment; the warmth and kindliness of his nature, in fact, made him friends in the process among many who had spent far longer in the company than he. In 1946, the issued share capital of the Renold and Coventry Chain Co. Ltd. was £1·2 million; in 1962, when his executive service ceased (although he remained, and remains, a director) it stood at £8·6 million.

Mr W. V. Foley, who succeeded Mr Bertoya as General Sales Manager in 1953, was appointed to the board in 1955. An engineer, he entered the Renold and Coventry Chain Co. Ltd. as a result of the 1930 merger, before which he had been employed by the former Coventry Chain Co. Ltd. Under the new regime he became European Technical Service Engineer until 1939, when he was appointed Inspection Manager at Coventry. Next, in 1944, he was made Sales Technical Manager. In 1946 he went to Sydney as South West Pacific Territorial Sales Manager, returning to England in 1951 to be Managing Director of Renold Chains Continental Ltd. He succeeded Mr Bertoya as General Sales Manager in 1953. In 1959 he was again overseas, this time as Executive Chairman of the Australian subsidiaries. He came back to rejoin the parent board in 1963. In common with others whose executive phase is over, he remained a director, retiring in 1968.

Mr L. J. Tolley, the present Group Managing Director, who heads the general management team now in office, was born in 1913, entered the radiators branch of Morris Motors Ltd. in 1929, and became an engineer. Ten years later he had risen to be Production Manager. From 1941, when he was 27, to 1953, he was General Manager of Nuffield Metal Products, thereafter becoming General Works Manager of Renold. In 1955 he was appointed a Renold director. Within the company he rapidly established his position through the respect commanded by his technical authority as well as by his qualities of character; together these rapidly gained the trust and confidence of those who carried out his instructions in the factories. As these instructions in a number of cases involved changed layouts, calculated partly to improve overall productivity and efficiency and partly to obtain the best results from newly-introduced techniques, natural conservatism might have found it difficult to keep pace with this dynamic newcomer, but his manifest devotion to the company, fairmindedness and enthusiasm were quickly communicated to others.

As Managing Director and Group Managing Director since 1962, his wide and ranging mind looks well beyond the Group's engineering requirements, formidable as are the demands that these make on any man, and embraces without apparent strain the whole field of business policies and principles. This is clear from a study of his paper entitled 'The International Operation of a Technical Product Company', presented to the Centre d'Etudes Industrielles at Geneva in March 1965.[1] In it he makes no pretence that business is to be judged other than by its profitability, itself a reflection of service rendered and of successful and humane management. His standpoint is made clear in the introduction, where he states that:

> 'Particularly striking failures in business assessments today are in the ability to appreciate that industrial undertakings have their own individual needs and policies, that the mixing of eggs and elephants on the evidence of good management is not a panacea of success, that initial price to one's customer is not the be-all and end-all of efficiency measurement, and that goodwill, tradition, service and technical excellence are still major factors in the continuing success of business, particularly in companies based upon technical products but also, I suspect, in more of the other companies than this modern world will officially allow.'

After a brief historical survey of the company, he discusses its main operating philosophies and policies. From this discussion emerges a pattern of management and control which he claims to be in some respects unique. He examines the pattern under the headings of the main operational functions—sales, technical application, research, personnel, finance and so on. The paper deals with Renold Chains Ltd. as such; since it was

[1] And subsequently to the Department of Management Studies, Birmingham University (November 1965), the British Institute of Management, N.W. Region (January 1966) and the Society of Investment Analysts (February 1966).

MR J. P. HALPIN

MR REX BATE

written the larger Group has evolved, embracing positive power transmission in the wider sense and symbolized in the name Renold Ltd., but the document stands as an assessment of the Renold industrial empire and its motivation. The temptation is strong to indulge in copious quotation, but 'The International Operation of a Technical Product Company' is better read as a whole and is therefore printed in full as Appendix I.

The other members of the team at the time this narrative ends were Messrs J. P. Halpin, Rex Bate, E. E. Hempsall and G. W. Taylor.

Mr J. P. Halpin's Renold career has been in some respects analogous to Mr Tully's. He, also, met with success in a profession unrelated to mechanical engineering, in his case the law, and in Government service and private business. Born in 1910, he was articled to, and subsequently worked in collaboration with, a solicitor who was also a prominent politician, the late Leslie Burgin, the Minister of Supply in the opening period of the 1939–45 war. He was appointed to the Treasury Solicitor's Department and was seconded in 1940 to the raw materials department of the Ministry of Supply. Wider responsibilities came to him, first as Assistant Secretary to the Ministry of Production, in which from 1942 to 1945 he served on the British Raw Materials Mission in Washington and secondly for two years following the war as an Assistant Secretary at the Board of Trade. He eventually turned from the Civil Service to industry, becoming Managing Director of Lewis Berger (Great Britain) Ltd., the paint manufacturers, before joining Renold Chains Ltd. in 1956 as General Administration Manager. He was appointed a Renold director in 1957.

Mr Rex Bate, the Sales Divisional Managing Director in the present team, is an electrical engineer who was formerly with ASEA of Sweden. After a period with the then Central Electricity Board he worked in Britain, Sweden and Germany for ASEA. During the war he served in the Royal Indian Engineers, attaining the rank of Lieut.-Colonel. In 1946 he joined the Brush Electrical Engineering Co. Ltd., of which he became Sales Director, until 1954, when he joined Enfield Cables Ltd.

as Sales Director, going to Renold four years later. He became Renold General Sales Manager in 1959, his appointment to the board following in 1960. He has been responsible for extensive reorganization of the company's selling force and agencies, the establishment of many selling subsidiaries, and the redeployment of the United Kingdom branches.

Mr E. E. Hempsall joined Renold in 1920, completing a special apprenticeship in 1928. His name first became prominent in the company because of his ability as an engineering designer, which he applied successfully to the production of special-purpose machine tools and other machinery. In this he continued in modern times and ways the traditions of machine development established in Hans Renold's day. He became Production Service Manager at Renold Works, Manchester, in 1950, subsequently holding managerial appointments at Cardiff and Coventry. Mr Hempsall later had charge of the new Burton-upon-Trent factory, after which he returned to Renold House in 1957 as Central Production Manager—an appointment no longer existing—until his promotion to General Works Manager in 1961 and to the board the following year, pending his assumption of the duties of Works Director. He retired in 1968, but remained on the board.

Five years later, the third general management team was necessarily involved in the reorganization of Group control when the name of the company became Renold Ltd.; this development has been described in Chapter 6.

The Group Managing Director, together with his general management colleagues in charge of finance, sales and works, bears the brunt. Nevertheless, they have the support and are subject to the critical judgment of the full board. Until recently the board membership included Mr C. T. Bayliss and Mr E. O. Liebert (who retired in 1967). Mr G. W. Taylor remained until 1968 as Managing Director of the Holroyd companies, one of the two largest and most important parts of the Group not of Renold origin, the other now being Crofts Engineers (Holdings). Mr Taylor's contribution to Holroyd has already been described in Chapter 3.

Whatever the shortcomings of all these men, past and

present, active and retired (and they have not been exempt from the common proneness of humankind to make mistakes), they have constituted successive executive teams that have risen to the level of events. They have also carried much more than the average burden of responsibility. Their successes have depended on the loyal support of a host of subordinates. But the qualities of these subordinates, high as they have been and are, have in turn rested on effective processes of selection devised by the leaders themselves.

The Group board's controlling powers are strengthened from other sources, the remaining members of the second general management team, Mr Bertoya and Mr Tully, and eminent men from beyond the engineering industry. Sir Thomas Robson, before his appointment to the Group chairmanship, was one of the latter, of whom other representatives are Sir Patrick Hamilton, Chairman of the Expanded Metal Co. Ltd., and Sir Leslie Robinson, formerly Second Secretary at the Board of Trade. Another was Sir Wilton Lee, who retired by rotation at the annual general meeting in August 1967; he did not seek re-election because of his appointment as first Chairman of the British Independent Steel Producers Association, the organization set up by the remaining privately-owned companies in the iron and steel industry following nationalization in that year. All these, with their records of work in official and other business, have brought in wide and energizing contributions of talent and wisdom from outside.

CHAPTER 9

THE GROUP PHILOSOPHY

From 1957 until 1967, when it was radically amended, each issue of the annual report and accounts of the company contained a brief statement describing the Renold business in standard terms. It was worded as follows:

'The business of the company is concerned with two main fields of engineering, namely, the mechanical transmission of power and the conveying and handling of material.

'The company's concern in power transmission is with the production of precision chains (mainly roller chains), the toothed wheels on which they run and various adjuncts such as clutches, couplings, cases for the enclosure and lubrication of drives and automatic chain tension adjusters.

'In the field of conveying, the company's contribution is the provision of an extensive range of conveyor and elevator chains, the attachments and the wheels on which the chains run.

'Transmission chains produced by the company range from chains weighing a few ounces a foot to some of over 60 lb. a foot used for example on marine diesel engines. Between these extremes fall the well-known cycle, motorcycle and automobile chains and a wide range of intermediate sizes for the varied transmission requirements of industry. The range of toothed wheels is equally extensive and provides, in conjunction with the chains, a wide selection of complete drives available from stock.

'The company's products, chains, wheels and accessories, are standardized and lend themselves to mass production, although special design and manufacture of all these is undertaken to meet the requirements of customers.

'The company's products are used in every industry where-ever there are wheels to be turned or materials to be conveyed mechanically and in every country in the world.'

These six short paragraphs make clear the company's connection with industry as a whole, and internationally. The relationship is permanent and sensitive. It is necessarily international, since the product or products are exported to or made in a large number of countries.

When the 1968 annual report appeared, the statement had been considerably edited to take account of the changes in the Group. The evolution that had occurred was recognized in the heading, which now read 'the business of the Renold Group', and the wording summed up the altered state of affairs. The two versions are worth comparing; the second ran as follows:

> 'The business of the group is the manufacture and distribution of products for power transmission and mechanical handling. To meet world demand, the Renold Group has an extremely wide product range comprised of the following major items: roller chains, wheels and pinions; conveying and elevating chains and wheels; gears; wormgear speed reducers; wormgear sets; spur, helical and spiral bevel gear-boxes, shaft mounted gear units; geared motors; variable speed drives, hydraulic and mechanical; couplings, clutches and brakes; sprag clutches for overrunning, indexing and backstopping; power transmission accessories.
>
> 'Virtually all requirements of the market are met by these products, many of which are standardized in design and thus lend themselves to mass production techniques. Special products are, however, designed and manufactured to suit the needs of individual customers.
>
> 'An international network of subsidiaries and agents covers almost every country in the world. Sales and service facilities are supported by stocks held in some 400 sales offices and depots where Group trained personnel are available to provide technical advice and assistance to customers.

'Additionally, the Group produces machine tools; helical rotors for air compressors; tablet-making and other machinery for the pharmaceutical, confectionery and similar industries; hammered steel forgings and copper base alloy castings and machined components.'

From this business arises the Group philosophy. It is based upon a robust faith in its British origins. The Group Managing Director, addressing the Renold Group international conference in 1965, made this very clear when he told 140 delegates from fifty countries:

'England is going through a period of economic stress. Our balance of payments position is maintained with borrowed money and many of you will know that when you are a borrower the lender insists on having something to say about your affairs. This economic stress is the result of an understandable desire for faster economic growth in this country. It has resulted in imports growing faster than exports, but it must not be forgotten that this very growth has helped many of the countries of the world in which you operate. International pressure has caused a slowing down in this policy. I have no doubt that as a result our balance of payments will improve, we shall continue to underwrite sterling trade and undertake world-wide obligations which many of our critics avoid like the plague. But I am equally certain that world trade will suffer and the slow-down in this country, which so many other countries believe is right for everyone except themselves, will reflect throughout the trading nations of the world.

'The greatest enemies of British economic strength appear to me to be the British themselves, and their Press in particular. I know of no other country which delights in exposing its weaknesses and faults and at the same time hiding its strengths. To correct this is our business, but I want to emphasize to you that these strengths exist.

'The critics of the United Kingdom concentrate upon our export difficulties, comparing our performance with others. Such comparisons are often meaningless. Our export performance over the years is by no means bad. These critics never mention British imports, which provide the means of livelihood to countless millions throughout the world, many in underdeveloped countries. They, of course, began to realize that these imports existed when a surcharge was imposed as a small contribution towards the avoidance of far worse consequences. The ever-open door of Britain's trade, despite her difficulties, is a magnificent example of international beliefs which the rest of the world must not forget if it requires the example to be maintained.

'I implore everyone to compare like with like, and this country will be seen as succeeding, not failing. If you wish for reassurance, remember our successes, think of Imperial Chemical Industries, the British motor industry, British steel, British agricultural equipment, chemicals, man-made fibres, construction equipment, and, if you like even Renold and Holroyd. Think of the workpeople of Renold. Go and see the workers at Holfos foundry and, having done so, be proud and remember that the prosperity of the company and yourselves depends upon the prosperity of Great Britain. The prosperity of Great Britain and her ability to play her part in the world has depended never more than now upon international confidence. Do your best to maintain that confidence. Destroy rumour, establish facts and only make proper comparisons.

'We meet at a critical time for Britain's economy. I have the confidence to be sure that we shall win through, that the new Group will go from strength to strength and that you will prosper with it.'

These positive statements reflect the convictions that animate Renold and enable the Group to bear industry's present troubles. A year later Mr Tolley wrote: 'The world uses sterling as its trading currency. . . . The going will be much tougher this year [the Group's financial year 1966–7] but we have a splendid world-wide organization and although greater effort than ever will be needed, we have the means to do well despite the difficulties.'

The figures since then justify his words. To take the latest, the report for the year ended on March 31, 1968, showed the Group profit on trading for the UK companies to be £4,369,000, compared with £3,328,000 for the year 1966–7. Group sales for the year 1967–8 (after deducting sales within the Group) amounted to £36,537,000. The severe restrictions imposed on the home economy in July 1966, and earlier, resulted in a lower level of home demand. Although this might be cushioned by exports and overseas business, the year's results would have been excellent, had they not reflected in some degree the impact of the Government's measures. The total Group trading profit was £5,456,000, compared with £4,211,000 for 1966–7.

Size for its own sake has never attracted Renold leaders; if it had, the Group would probably have developed into a strictly holding company financially controlling many disparate enterprises. Instead it has grown into a British and international manufacturing business not, as was once said of the growth of the British Empire, in a fit of absence of mind, but in consequence of the world's demand for engineering products. This growth took first the possibilities of precision chain only as its chosen field. But inevitably there followed chain's immediate accessories and ancillaries, then clutches, and now wormgear drives and other means of positive power transmission. It is already apparent that the logic of events is causing the movement into fields beyond chain to go still farther. Significantly, the connection between these differing engineering products is organic; although each can be used separately, it can also be used in combination with the others. Herein there may well lie the reason for the Group's growing strength, given a system to develop its full potential, and policies to make the system effective. It should not be forgotten, however, that in addition to successful invention, design, production and marketing of things the world has been eager to utilize, there has been another contributor to Group size and geographical distribution: Renold and Renold subsidiaries enjoy no monopoly; the invigorating stimulus of competition is always present in other power transmission products; these and competing chain manufacturers and merchants present a constant challenge.

RENOLD HOUSE

'Whatever thy hand find to do, do it with thy might', but Hans Renold wished the 'might' to be tempered with love of one's neighbour. The letter and spirit of constitutional management clearly show this persistent conviction in relationships with the workpeople. 'The right and proper way', to use Mr Tolley's phrase, in which to deal not only with them but with all who come into contact with Renold Ltd. is dominant. This is clear, even in comparatively small matters, like the purchase of land at Burnage Lane, Didsbury, in 1961, in order to provide as a public amenity a recreational garden adjacent to Renold Works, Manchester. The list of sums disbursed in aid of good works, educational, medical and in other ways benevolent, provides more evidence in this respect; there are over 200 organizations to which Renold House and the chain division manufacturing establishments alone make annual donations or pay annual subscriptions. In at least half of them it is impossible to trace the remotest advantage to the givers, at any rate in this world, but the gifts illumine the strong sense of communal responsibility rooted in the Group philosophy. In a different field, the general management team, and others, take their full part in work done for industry through service on professional, technical and cultural bodies. This may help the Group, but it also means heavy demands upon time and effort to the benefit of industrialists at large, including competitors, and is another facet of the way the Group discharges its duty to the community.

An independence of spirit pervades Group thinking. Willingness among workpeople to accept trade union membership, wherever it may apply, is written into the rules giving effect to constitutional management. But management reserves its own freedom on the other side of the table, and its right to go its own way, picking and choosing as suits it where employers' organizations are concerned. It is similarly independent of general fashion or custom in employing no advertising agents and no public relations organizations; Group publicity has a crisp professionalism of its own, and its public relations are part and parcel of its normal conduct internally and externally.

A thought-out set of principles, a state of mind, a code of

conduct, govern Group policies and decisions. The philosophy motivates the thinking and action of the Group's leaders, extends to the machine shop and the office, sustains and animates the sales force. It takes account of the many different people by whom and for whom it exists, in the clear light shed by a strong sense of moral obligation. The corporate spirit in the Group stems from a traditional attitude to humankind which has been carefully preserved. It is, in fact, timeless, and a powerful ally in technical progress and service to the customer. It would be fallacious to suppose that it has invariably been taken to heart by every human soul concerned with it. Such sentiment is an aspect of the nature of each person with all his or her proneness to mistakes, misunderstandings, and failure. The possibility, the inevitability even, of setbacks and difficulties is not denied, but the ethos, the habitual character and disposition, of the Group allows for this fact, and seeks to mitigate or prevent their ill effects by sensible action in good time. A perception of industrial and commercial evolution as something to be derived from and guided by ethical concepts has characterized Renold policy throughout the period described in this book and its predecessor. This is more than emotional indulgence. It has rewarded the enterprise with success and is making a moral and material contribution to posterity.

* * *

The present account has told mainly of an era in Renold history of which the outstanding marks were the merger with Holroyd, the new Group organization, and the death of Sir Charles Renold. The first step in the succeeding phase is the acquisition by Renold of another famous positive power transmission company, Crofts Engineers (Holdings) Ltd., of Bradford. The Crofts organization produces and markets nearly 100 kinds of transmission and other related products, including gears, clutches, couplings and drives, and brings into Renold fifteen sales, or sales and stockholding, branches in the United Kingdom, one in Dublin, and four associated companies, two in Canada, one in South Africa, and another in the United States,

MR E. E. HEMPSALL

MR G. W. TAYLOR

together with a world-wide network of agents and stockists. With what has gone before, this makes Renold internationally the largest positive power transmission grouping, in the specialized sense in which the Group uses the phrase 'positive power transmission'. Renold now take a new stride forward into yet wider regions of application for positive drives, the far-reaching effects of which have yet to be seen, and must await a record in other pages than these.

APPENDIX I

THE INTERNATIONAL OPERATION OF A TECHNICAL PRODUCT COMPANY

by L. J. Tolley, Managing Director, Renold Chains Ltd.
(Note: Mr Tolley is now Group Managing Director of Renold Ltd.)

As presented to the Centre d'Etudes Industrielles, Geneva, March 23, 1965; Birmingham University, Dept. of Management Studies, November 25, 1965; British Institute of Management NW Region, January 20, 1966, and The Society of Investment Analysts, February 1966.

My paper is intended to be somewhat of a tour d'horizon *of Renold business policies and principles. For a company to hold strong views about the way its business is conducted can have many advantages. To Renold not the least of them is that it represents the logical road to profit and financial success.*

Particularly striking failures in business assessments today are in the ability to appreciate that industrial undertakings have their own individual needs and policies, that the mixing of eggs and elephants on the evidence of good management is not a panacea of success, that initial price to one's customer is not the be-all and end-all of efficiency measurement, and that goodwill, tradition, service and technical excellence are still major factors in the continuing success of business, particularly in companies based upon technical products but also, I suspect, in more of the other companies than this modern world will officially allow.

It is this field of industrial management which I propose to emphasize to you in my paper today.

I must start with a brief review of my company's history, its products and its current structure, for it is necessary to understand a little of this in order to appreciate some of the business principles which I propose to outline a little later.

Renold Chains Ltd. started its life in Manchester, England, in 1879. It was founded by a Swiss engineer named Hans Renold, born in Aarau, who had made his way to England via Paris a little earlier to work in Lancashire on the design of textile machinery.

The Company's initial capital was £300, and it had a very fortunate start. It made textile chains, and because of this, when Starley invented the safety bicycle in Coventry, England, in 1880 he approached Hans Renold for assistance in solving his chain drive problems. The outcome of this was the Hans Renold invention and patent of the bush roller chain which is basically the same product made throughout the world by the mightier Renold company today

Now Hans Renold was an intuitive engineer of great vision who possessed some very determined principles upon how a company of the type he envisaged must develop. He knew that cycle chain was just a beginning, that roller chain could become the most efficient means of transmitting power ever known, but only if it acquired a high reputation through quality and consistency. He therefore conditioned his whole approach to business upon QUALITY, SERVICE, TECHNICAL EXCELLENCE and APPLICATIONAL KNOW-HOW. He had the foresight to appreciate that his customers could ruin this reputation by wrongly applying his products, and even in the very early days there is a Renold story of a powerful businessman from the English Midlands who wished to buy many thousands of feet of roller chain from Hans Renold. He was politely asked how he proposed to use it. The businessman replied angrily that this was his business, he would use it—Renold would supply it. He was quietly told by Hans Renold, 'If I do not know how my chain is to be used, I do not allow my chain to be used.' To a struggling young engineer the refusal of this kind of business must have been extremely courageous, but it was this type of courage which, followed throughout the years, has enabled the company to say that British Standard roller chain is correctly applied throughout the world and that it can be replaced and renewed in any application by a standard, catalogued product without fear of failure or complication.

Over the years Renold Chains has grown as a result of its own development and by acquisition and merger, and, including its most recent acquisition, John Holroyd and Co. Ltd., which will be the subject of a short reference later, it now employs over 11,000 people. Its Stock Exchange market value is now £29,000,000, with net assets currently valued at £26,000,000, and after depreciating its assets by over £1,000,000 annually it has a trading profit before taxes, loan interest and dividends of between £3,000,000 and £4,000,000 per year. It has some twelve factories in the United Kingdom and also manufactures in France, Germany, Australia, Canada, Spain and Argentina. It has its own selling subsidiaries in nine overseas countries and actually sells its products in over one hundred countries throughout the world. This large business empire still maintains many of the original edicts of Hans Renold, we hope suitably modernized and brought up to date.

I am now going to tell you something of our main operating philosophies and policies, from which will emerge a pattern of management and company control which, although in some respects unique, I hope you will agree is appropriate to a technical product company. It is in any case successful.

SELLING

Let us start with our selling philosophies and policies. They are:

(1) To believe in the technical advantages of precision chain and the vastness of the legitimate technical and technological field for which it caters.
(2) To believe in quality and technical ability to perform.
(3) To believe in the benefits of quantity production.
(4) To believe in the benefits of standardization, variety control and minimization of specials.
(5) To believe in maximum service from stock.
(6) To believe in the education of customers in these same beliefs and in the propagation of sound chain applicational practice, IF NECESSARY AT THE COST OF IMMEDIATE BUSINESS.
(7) To believe fanatically in the selling of Renold branded goods and to guard jealously the Renold mark.
(8) To believe wholeheartedly in orderly channels of distribution and the sanctity of discount scales.
(9) To believe in maximum direct contact with the ultimate user.

PRICE

With selling we must of course associate the policy of price. A technical product needs a price but it is important to remember that INITIAL price is not the be-all and end-all of selling. With a technical product of the kind with which we are concerned, VALUE is of prime importance and value is measured by performance, wearing qualities, efficiency and service as well as price. Price is but a yardstick against which we measure value. Anybody can make a product cheap—it's easy—you just leave out a lot of things, but the talent for making a lower price for a quality product is much rarer. Any technical product company which operates by initial price alone is unlikely to be a continuing business success. Manufacturers are successful in business because the features and services of their product 'measure up', and it is because they are successful that they can take the lead in research, development and production and can carry the really high costs of continually producing and marketing new and better products.

SERVICE

Adequate service and availability stem from standardization and the ability to have supplies in the right place at the right time. The major contribution to service of a technical product throughout the world depends upon adequate stocks of that product throughout the world. In this computer age it is fashionable to hear the cry of 'over-investment in inventory' and 'idle money'. Again these financial requirements do not always apply to the servicing of technical products. Theoretical inventory, based upon history rather than need and possibility, will lose more sales of a technical product than any other known restriction. This is not of course to say that stocks must not be properly controlled. The financing of such stocks, particularly overseas, is a problem, but when they concern products which are standard and scarcely ever obsolete then the risk is small and the risk of non-availability and its consequences so much greater.

MANUFACTURING AND QUALITY

Our manufacturing policy insists that we not only maintain but continually improve our efficiency. Efficiency, like quality, is a state of mind but it is based upon technical progress. Continual technical probing and investigation into our manufacturing methods is considered to be essential. We are never satisfied with what we are doing—we know there is always a better way. Nowhere else in the chain industry in the world will you see machines and processes as modern and efficient as you will see in the factories of Renold Chains. Fine manufacturing equipment is a most important factor in product quality, that part of our manufacturing policy of which every Renold employee is justly proud. There ARE a few people in the world—I'm sure they all work for Renold—who can adequately and properly SPECIFY a quality chain, but the production of a finished product of consistently high quality depends upon specification plus manufacturing control. Quality control is pre-eminent in all Renold factories. It is costly and almost awe-inspiring. But never believe people who say they can achieve quality on a shoestring budget. It is necessary to pay for quality if you want it, and we do.

TECHNICAL APPLICATION

This aspect of our operating policies concerns not only technical services to customers and the evolution of new applications but the protection of our product reputation against user malpractice. Our aim is not just to ensure that chain is used but that it is used correctly. To this end all our salesmen throughout the world are engineers trained in the correct selection and usage of chain drives, supervised by Territorial Sales Managers and supported by home-based Applicational Engineers who can give immediate advice and specification for any unusual or difficult application. Our technical literature is designed to make chain selection for normal application easy but to draw a thick red line between the normal and the abnormal. Of course we have difficult customers who insist they know best, but our powers of persuasion are substantial and our results are available to prove our advice when necessary to their and our own satisfaction. We are, however, insistent in the overall interests of chain driving that only the best practices and design are permissible.

RESEARCH AND DEVELOPMENT

Our research and development policy has as its aim to cover fundamental and applied research upon all aspects of our products including design, manufacture and application, and to place ourselves in a position based upon the knowledge provided from our own resources to stand four-square behind any technical pronouncement made in connection with our products. The resources of our Research and Development Department are indeed adequate and our people first class. We can carry out all research science, both fundamental and applied. Our chemical and metallurgical laboratories provide aid to our steelmakers themselves. Our fatigue laboratories have outstanding equipment which can cover our whole product range. And our testing equipment for all our products gives us speedy answers to our own problems and helps to assert our competitive position. We have an internal combustion engine test house which is the envy of many automobile manufacturers and which has made us the leading experts in the world in engine camshaft drive design. We know that research and development is our investment in the future.

EMPLOYMENT POLICY

Our basic policy in this field requires that those we employ shall enjoy working for us, shall wish to continue working for us and shall derive satisfaction from doing so. We believe that all people want to work and to work fairly, and we accept that it is our responsibility to provide conditions in which we can reasonably expect this to happen.

Joint consultation at all levels plays an important part in our employee relationships and, apart from continuing discussion upon various aspects of employment and business, management is charged with the responsibility of ensuring that all decisions which are likely to affect people are adequately discussed whenever possible before they are made with such people. Communications, again, at all levels are most important and formal procedures exist to ensure that the content and timing of any dissemination of information is adequately controlled throughout the chain of command.

We believe in, and encourage, trades unions. It is proper and indeed essential that employees of a large organization should be able in an organized and representative way to negotiate and bargain in all matters which affect them. We have never had cause to regret our faith in organized labour.

But all these sound ideas on employment and employee relationships need, in my view, an important catalyst—it is 'opportunity'. To everyone there must be given opportunities in one's work for improvement, advancement and greater satisfaction. Without this no employment policy can really be successful. Opportunities, however, cannot be left to the filling of dead men's shoes, they must be created; not, though, as I fear sometimes happens in some of our larger organizations, by giving meaningless jobs high-sounding titles but by providing really worthwhile jobs in a growing and ever-expanding international organization. Growth is the key word not only for profits and business success but for the maintenance of successful employment policies. Employment policies and growth lead us quite naturally to the training and development of people.

A growing international company such as Renold always needs people. Many years ago Hans Renold said to his son, our present Chairman, 'It is not our job to make chains, our job is to make people—they will make the chains for us.' But, apart from needing people, a widely based international company can within its own resources also provide unique means of training the people it wants. There are, of course, at Renold many ways of introducing and initially training people for life in a technical product company.

We have the normal apprentice training programmes for young people leaving school at the age of sixteen. These programmes provide adequate supplies of skilled tradesmen type labour but obviously today produce less and less in the way of supervisors or leaders, although even now a young man will come to the top through this method, and we ensure that he has the opportunity to do so. We also have schemes to cover young people leaving school from the sixth form, not going to university but taking their diploma in technology through the sandwich course method; and of course we have university graduates entering industry directly through our graudate training schemes.

We believe in training for management and support courses and universities which are active in this field. We particularly welcome in England the new schools of management education.

Initial industrial training is, of course, important but we believe that most important of all is to select, encourage and train those young men with leadership potential. To this end annual reviews of those with such potential take place and action is taken to broaden their knowledge and experience and to increase their overall abilities. This action takes the form of training overseas, situation training, planned promotion, special projects, etc., and it is at this time that our international resources are able to be used to their full value. With all this training we are not afraid to engage the right people from outside the organization as appropriate and necessary, and our policy also calls for the injection into our executive stream of a number of nationals other than British who after a period of training in the United Kingdom are employed to staff our overseas organization in many ways.

No training and development scheme is perfect, so much depends upon people. But with the policies outlined I am able to say that we have at all times a management and senior supervisory structure capable of undertaking expanded and additional activities. And what company, determined upon a policy of growth, could ask for more?

FINANCING THE COMPANY—HOME AND OVERSEAS

As you know, Renold Chains is a public limited company. The capital structure of the parent company consists of a relatively small number of preference shares and ordinary shares at present issued numbering 13,641,448 at par value of twenty shillings and a current market value of over forty shillings. These represent the bulk of finance, augmented more recently by three million pounds of debenture loan stock maturing in twenty-five to thirty years. The assets of the Company are revalued at appropriate times to ensure that their true value is indicated in the accounts, and depreciation policy is related to this true value, thereby ensuring adequate cash flow for asset replacement. It is necessary to be aware of this type of financial structure to understand and appreciate the policies which ensure that new finance for expansion and growth is available when needed. The United Kingdom today is a highly-taxed society. Company profits have always been a convenient target for such taxation and the National Budget in April 1964 was no exception in this respect, nor indeed I imagine will be that of 1965.

Distribution of profits in the form of dividends or extra stock issues therefore needs a fine balance of judgment between the dividend income, the stock capital gain and the profit cover against economic eventualities. Investors in the UK, unlike some other countries, expect a dividend to be regular. They of course like increases but expect such increases to be maintained. It is therefore the profit cover for such dividends which tends to be the variable, and with Renold Chains it has varied from 1·5 to 2 times cover over the years. Bearing in mind that depreciation is on revalued assets this is usually looked upon in financial circles as very satisfactory. Suffice to say that raising finance in any form for any expansion of our activities has never presented problems. The difficulty is always the timing of such financing so that the delay between the provision of finance for new ventures and the time when they are earning satisfactory profits is not too long.

Financing at home, therefore, is relatively easy. But financing overseas is another matter. The major difficulty springs from the United Kingdom's controls on overseas investment due to balance of payment difficulties and our own determined policy to be in control of all our overseas operations and investments. This requirement in our overseas policy is again associated with the fact that we are a technical product company. We are not just investing money overseas, we are introducing the manufacture of a quality technical product into other countries and we have to do so in a way which protects our quality marks and our goodwill. Such attributes built up over many years can be squandered overnight by precipitate action and thoughtless decision. We also believe that the development of these product manufactures must be determined upon an international basis, not a national one, and when we have partners or even minorities in other countries it is almost impossible for international judgments to dominate those of national character. Overseas finance, therefore, comes from three major sources—the initial investment from the United Kingdom parent company after much discussion with the Bank of England; ploughback of profits—again with increasing difficulty due to overseas investments control; and local loans.

So far we have managed very well, but it is my certain conviction that availability of international finance is the greatest restriction on the overseas development of an international company, and I see this situation becoming worse as United States companies are further restricted in their overseas investments, because they certainly will be as the dollar situation weakens.

FINANCIAL CONTROL

I do not intend to bore you in this section with facts and figures, but merely to state quite briefly our principles of financial control.

We believe in budgetary control covering forward and annual

sales budgets, trading budgets, finance budgets and capital expenditure budgets. Budgets, however, are not documents authorizing expenditure; they are for approval, limitation and guidance. Actual expenditure must follow the pattern of activity. Our control documents take the form of periodic control accounts—which are different from financial accounts. We believe that the latter are so often confused by the need to comply with fiscal and taxation laws that they are inadequate for the purpose of judging the progress and performance of the company throughout the year.

We insist upon identical financial accounting and control throughout all our companies, at home and overseas, we can always adjust for national requirements, and only in this way can we be sure that results are correctly presented and comparisons correctly made.

CONCLUSION

And so those are the various facets of the operational control of an international technical company product. It is a rather special company and obviously its policies and philosophies cannot universally apply. It has itself considered whether in this modern world it can continue being a company relying upon a single type of technical product and whether its future must be somewhat different from its past. The conclusion from this self-examination has been that its policies and philosophies are correct but that they can be applied over a wider field with all-round benefit. The choice of this field is important if the present success and goodwill resulting from the tradition of a single technical product company is not to be endangered. The choice was therefore made a few years ago that Renold would enter the field of other transmission products which as far as possible should be complementary to the existing product rather than in competition with it. In this way not only can we achieve greater economy and efficiency in our own and the expanded fields but we can give greater service and satisfaction to our transmission customers. A start has been made with the acquisition of John Holroyd and Co. Ltd. whose worm gears are a perfect complementary product. We are currently engaged in the rearrangements resulting from this acquisition, but I have little doubt that in a few years' time Renold will have established a reputation associated with positive drive transmission equipment which will equal that it already possesses in the field of precision roller chains. Allow me now to summarize the policies of Renold Chains Ltd.

A TECHNICAL PRODUCT COMPANY

Because of its belief in quality, service and applicational know-how it has upon foundations of tradition and goodwill built up an industrial empire which possesses an international reputation for its business ethics, for carrying out its commitments and for operating in the right and proper way. All these things represent the life blood of an international technical product company if its success is to be a continuing and developing one. It cannot substitute opportunistic business methods for those upon which the goodwill of its product and its business is based. In following these lines such a business not only produces successful results but it creates and sustains good people, and this is perhaps one of the best advertisements for its policies.

We at Renold believe in the right and proper way in business, not only in the technical, engineering and financial fields, but in the employment of people and the associating with and servicing of other people and in business ethics, business moralities and our very way of life. We believe that if we insist upon doing things the right and proper way others can only follow. This all becomes possible in a technical product company when there is quality in the product and quality in the people. We also believe that these things are possible in other business, as we indeed believe that this is the role for business in the future progressive democratic societies. Those who join us and believe with us in these philosophies will not only enrich themselves but will make their contribution to the enrichment of the world.

APPENDIX II GROUP ORGANIZATION

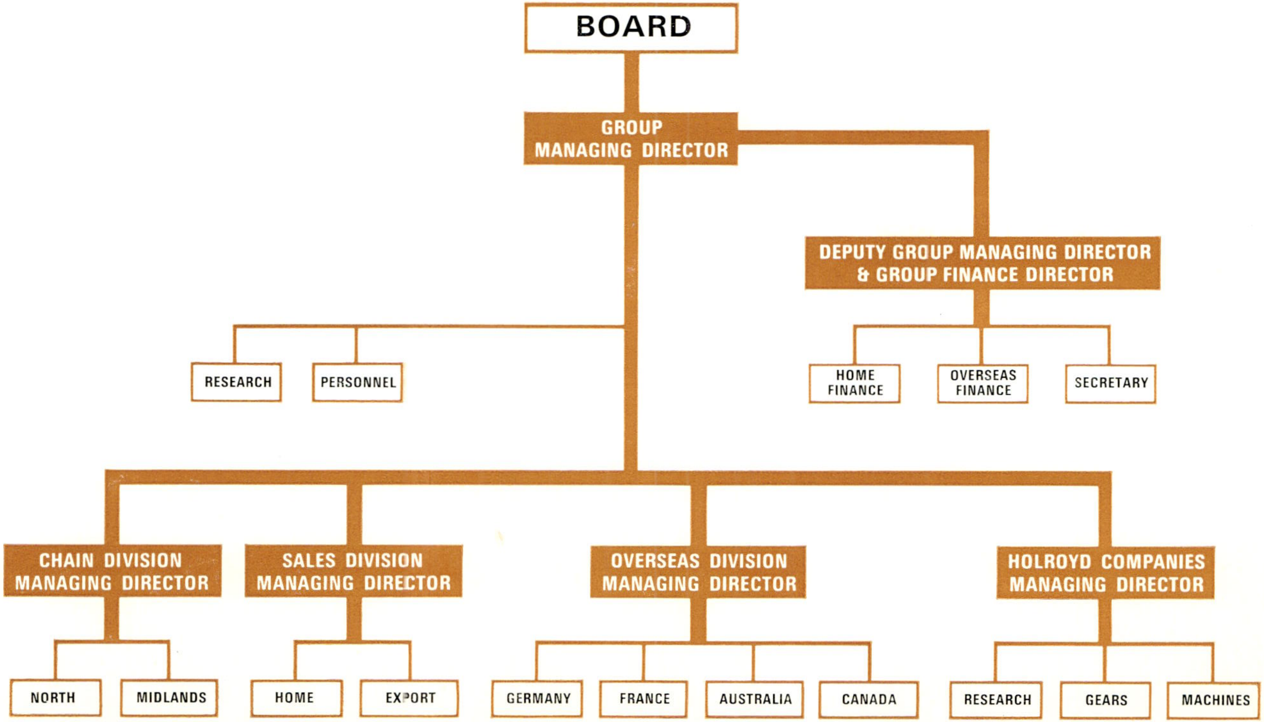

Since this chart was prepared, the acquisition by Renold Ltd. of Crofts Engineers (Holdings) Ltd. has led to the creation, in place of the 'Holroyd Companies', of a new division known as the 'Holroyd and Crofts Division'.

INDEX

Index

GEORGE ALLEN & UNWIN LTD

40 Museum Street, London WC1
Telephone: 01–405 8577

Park Lane, Hemel Hempstead, Hertfordshire
Sales, Distribution and Accounts Departments
Telephone: 0442 2361/2

Auckland: PO Box 36013, Northcote Central N4
Barbados: PO Box 222, Bridgetown
Bombay: 15 Graham Road, Ballard Estate, Bombay 1
Buenos Aires: Escritorio 454–459, Florida 165
Beirut: Deeb Building, Jeanne d'Arc Street
Calcutta: 17 Chittaranjan Avenue, Calcutta 13
Cape Town: 68 Shortmarket Street
Hong Kong: 105 Wing On Mansion, 26 Hancow Road, Kowloon
Ibadan: PO Box 62
Karachi: Karachi Chambers, McLeod Road
Mexico: Villalongin 32, Mexico 5, DF
Nairobi: PO Box 30583
New Delhi: 13–14 Asaf Ali Road, New Delhi 1
Toronto: 81 Curlew Drive, Don Mills
Philippines: PO Box 157, Quezon City, D-502
Rio de Janeiro: Caixa Postal 2537-Zc-00
Singapore: 36c Prinsep Street, Singapore 7
Sydney, NSW: Bradbury House, 55 York Street
Tokyo: PO Box 26, Kamata